NANOTECHNOLOGY SCIENCE AND TECHNOLOGY

NANOSCALE-ARRANGED SYSTEMS FOR NANOTECHNOLOGY

NANOTECHNOLOGY SCIENCE AND TECHNOLOGY

Additional books in this series can be found on Nova's website under the Series tab.

Additional e-books in this series can be found on Nova's website under the e-book tab.

NANOTECHNOLOGY SCIENCE AND TECHNOLOGY

NANOSCALE-ARRANGED SYSTEMS FOR NANOTECHNOLOGY

KIRILL L. LEVINE
EDITOR

nova publishers
New York

Library of Congress Cataloging-in-Publication Data

Nanoscale-arranged systems for nanotechnology / editors, Kirill L. Levine, General and Technical Physics, National Mineral Resources University, St. Petersburg, Russia.
pages cm. -- (Nanotechnology science and technology)
Includes index.
ISBN 978-1-63482-353-1 (hardcover)
1. Nanostructured materials. I. Levine, Kirill L.
TA418.9.N35N3443 2015
620'.5--dc23

2015006805

Published by Nova Science Publishers, Inc. † New York

CONTENTS

PREFACE

In this research compendium volume of Smart Nanocomposites, various aspects related to nanoscale-arranged systems are discussed. Synthesis, characterization, modeling, and examples of practical usage are in the list of topics brought up in this book. In the discussion are unusual phase transition properties of some oxides, transport properties, electrochemical characterization methods, and optical, electronic, and chemo-sorption properties of nano-srtuctured systems synthesized by various approaches.

NEW MATERIALS

Vanadium dioxide was brought to the focus of research in numerous laboratories because of its intriguing phase transiotion behavior. Optical information switching in vanadium dioxide and iodine nanoparticles was stidied by Khanin and co-workers. Phase transition from metal to semiconductor studied by Ilinsky and co-workers was discussed in terms of electron-related correlation effects responsible for observing phenomena's. Applications in a variety of sensors, super-broad-band transistors, and superfast optical limiters are suggested for those types of structures. Chemosorption in Vanadium dioxide films was described by Tutov with co-workers.

Iodine transport is a tool not only for transferring metals onto surface, but also for implanting them into material, as was shown by Bogdanov. This method also can be used for acceleration of exchange reactions with metals participation.

Low-dimensional oxide systems are another example of nanoscale arranged systems of interest for nanotechnology. Surface lamination and atomic layer deposition methods were described by Ezshovskii. Nanodimentional effects in ponderomotoric interactions, such enhancement in local field strength are discussed in works of Moshnikov and Maksimov.

Electron beam processing was used for the creation of poly(vinylacetate)/graphite nanocomposite with enhanced vibro-absorbing properties, as shown by Mjaking and co-workers.

Continued work on lithium borosilicate glasses coatings with metallic cations was reported by Kashif with co-workers with the emphasis on Infrared spectroscopy as a technique of choice for characterization.

Nano-sized polymer additives for water-based drilling fluids were reported by Manea. Rheological properties of those fluids were evaluated under severe (high pressures, high temperatures) simulated drilling conditions.

Devices for Medical Application

Medical aspects, such as targeted drug delivery and photodynamic therapy of cancer, are reflected in publications of Spivak and Vlasenko.

Synthetic Methods

Among synthetic methods used for nanocomposites preparation, sol-gel is a popular laboratory approach since it is reproducible, cost-effective and diverse, as discussed in paper by Abrashova and co-workers. Precipitation with the help of the surfactant was used to prepare quantum dots in colloidal form. Synthesized nanocrystals were suggested for usage in chemical detection in paper by Mazing and co-workers.

Electrochemical methods remain leaders as a versatile and flexible technique to prepare nanoobjects. By anodic local oxidation reported by Maximov titanium oxide semiconductor was synthesized, while electrochemical anodization of aluminum foil was reported for preparation of anodic porous alumina by Muratova.

Modelling

Thermodynamical aspects (enthalpy of the formation of intermetallic cadmium tin nanoparticles) were analyzed by Barbin. Approaches to nanostructures characterization by impedance methods were provided in communication by Bezmaternykh and others.

Quantum chemical calculations for titanium oxide nanostructures were reported by Drozdov, while dispersion forces in nanoscale structures of metal-oxide semiconductor were modeled by Lifshitz approach by the group of Fedortsov. Approaches in band modeling of layered semiconductors were addressed by Nemov.

In: Nanoscale-Arranged Systems for Nanotechnology
Editors: Kirill L. Levine
ISBN: 978-1-63482-353-1

Chapter 1

STRUCTURAL STUDY OF LITHIUM BOROSILICATE GLASSES CONTAINING BOTH IRON AND NICKEL CATIONS

I. Kashif, S. M. Salem, H. Farouk, A. G. Mostafa, Sh. Salem and A. M. Sanad
Physics Department, Faculty of Science, Al-Azhar University, Cairo, Egypt

ABSTRACT

Glass batches were prepared according to the molecular formula:

$$[20\ Li_2O - 31B_2O_3 - 34SiO_2 - (15 - x)\ NiO - xFe_2O_3]$$

where x =0.0, 2.5, 5, 7.5, 10 and 12.5 mol%. In addition a base sample free from Ni and Fe was melted (35 Li_2O - $31B_2O_3$ - $34SiO_2$).

Both infrared and Mössbauer spectroscopy were used to investigate the structural changes caused by the replacement of nickel by iron cations. The infrared spectra were measured over a continuous spectral range (400 to 2500 cm^{-1}). They show that the replacement of Lithia by nickel oxide decreases the non-bridging oxygen content. Also, the replacement of nickel by iron decreases the concentration of non-bridging oxygens up to 7.5 mol% and then increases it.

From the values of the Mössbauer parameters, all the ferrous cations were found to be in octahedral coordination, while all the ferric ions occupied tetrahedral coordination sites.

The ratio $\{Fe^{3+}/\sum Fe_{total}\}$ increased on replacing nickel by iron up to 7.5 mol % and then became constant. The density increased by replacing nickel by iron up to 7.5 mol% and then decreased.

INTRODUCTION

Infrared and Mössbauer spectroscopy have been extensively employed over the years to investigate the structure of glasses. The alkali - B_2O_3 - SiO_2 ternary glass systems, in particular, have been the subject of numerous infrared studies due to their technological importance. These glasses are used for many applications such as optical glasses, nuclear waste materials and in the electronics industry. In alkali borosilicate glasses with relatively low alkali oxide content, alkali oxide is believed to associate itself with B_2O_3 alone forming an alkali borate phase while SiO_2 forms a silica phase [1-4]. The infrared spectra of these glasses show that alkali oxide is incorporated in the glass to form borate groups such as those found in binary borate glasses [5]. These groups consist of BO_3 and BO_4 units without non-bridging oxygen. The infrared spectra of alkali silicate glasses display a main band at about 1000 cm^{-1}, attributed to the formation of SiO_4 tetrahedra with non- bridging oxygens [6]. Also, the infrared spectra of Na_2O - B_2O_3 - SiO_2 and Al_2O_3 -Na_2O - B_2O_3 - SiO_2 glasses showed bands in the region 1000 – 1120 cm^{-1} arising from the contributions of silicate and borate groups, depending on the concentration of SiO_2 and B_2O_3 [7]. The strongest B - O band in the infrared spectra of Li_2O - B_2O_3 - SiO_2 glasses exists at 1270 cm^{-1}, while for the Si – O band it is at 1060 cm^{-1} [8]. A B – O – Si band was found at 440 cm^{-1} with an intensity mainly a function of composition [9].

The effect of replacing MnO_2 by Fe_2O_3 on the oxidation states of iron in some lithium borosilicate glasses was studied using Mössbauer, infrared and DTA measurements. The Mössbauer spectra showed that the iron ions appeared in the ferric state in both tetrahedral and octahedral coordination and the ratio between the numbers of iron ions in both coordination states did not change with increasing MnO_2 content [10, 11].

The present paper examines the structure of lithium borosilicate glasses containing both iron and nickel cations, in addition to a base sample free from nickel and irons using infrared and Mössbauer spectroscopies, which are valuable tools for the study of amorphous materials. Also, the densities were measured and calculated.

EXPERIMENTAL WORK

Chemically pure oxides were used to prepare the glass batches. The obtained batches, after complete mixing, were melted in platinum crucibles using an electric furnace type (VAF 15/10) LENTON thermal designs at 1200°C ± 20°C for two hours. The melts were stirred during melting to ensure complete homogeneity. Then the melt was poured in air on a stain-less steel plate, at room temperature. The samples were divided into two parts. One was powdered for the Mössbauer and IR spectroscopy measurements. The other part was used in the solid form for density measurements. The samples were examined using a Philips Analytical X-Ray diffraction system, type PW 3710 with a Cu tube anode. XRD showed no evidence of crystallinity in the quenched glass samples.

The infrared spectra were measured using the KBr disc method and recorded using a JASCO FT/IR 5300 infrared spectrophotometer in the region between 400 and 2500 cm^{-1}.

The Mössbauer spectra of 70mg/ cm^2 samples were measured using a constant acceleration spectrometer and a 1.85 GBq, Co^{57} source diffused in chromium matrix. A least square-fitting program based on the line shape distribution was used in order to determine the

Mössbauer parameters. The density of the glass samples was measured at room temperature by the Archimede`s technique, with an accuracy up to ±0.0001 g/ cm^3. The samples were weighed in air W_a and in a liquid W_l with a known density such as toluene (d_t). Then the density of the samples was calculated using the following formula:

$$d = \{ W_a / (W_a - W_l)\} \times d_t$$

where d is the density of the sample and d_t = 0.8655 g/cm^3.

RESULTS

Infrared spectra for the glasses melted have been measured and recorded in the region from 400 to 2500 cm^{-1}, to obtain information about changes in the vibration spectra for lithium borosilicate glasses containing different concentrations of NiO and Fe_2O_3.

Figure (1) shows the IR spectra of the investigated glass samples. The spectrum of the lithium borosilicate glass sample (free from nickel and iron) displayed seven absorption bands, at 1420, 1030, 700, 520, 500, 460 and 440 cm^{-1}. And the assignment of the absorption bands detected is summarized in Table 1.

Table 1. The assignment of the absorption bands

Fe – Ni Content	0 - 0	0 - 15	2.5 - 12.5	5 - 10	7.5 - 7.5	10 - 5	12.5 - 2.5	15 - 0**
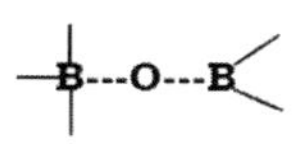	1420	1400	1400	1400	1400	1400	1400	1420
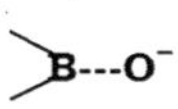 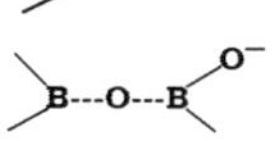	1030	1020	1000	1030	1020	1030	1000	1020
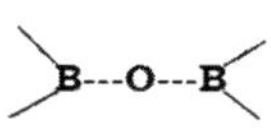	700	700	700	700	700	700	700	740-715
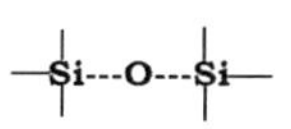	520	--	540	530	580	---	---	550-500
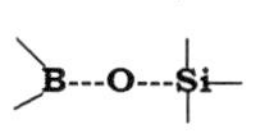	460 440	460 430 410	480	480	480	480	470	475- 460

** I. Kashif et al. Physics and Chemistry of Glasses 29(1988)72.

When replacing 15mol% Li_2O by 15mol% NiO it can be observed that: the two bands at 700 and 460 cm^{-1} are unaffected, and the absorption band at 520 cm^{-1} disappeared. A new absorption band appeared at 410 cm^{-1}, while the bands at 1420, 1030 and 440 cm^{-1} shifted to lower frequencies at 1400, 1020 and 430 cm^{-1} respectively.

The molar replacement of NiO by Fe_2O_3 showed that: the absorption bands at 700 and 1400 cm^{-1} were unaffected and appeared in all the investigated glass samples, while the absorption bands at 1020 and 430 cm^{-1} were randomly affected.

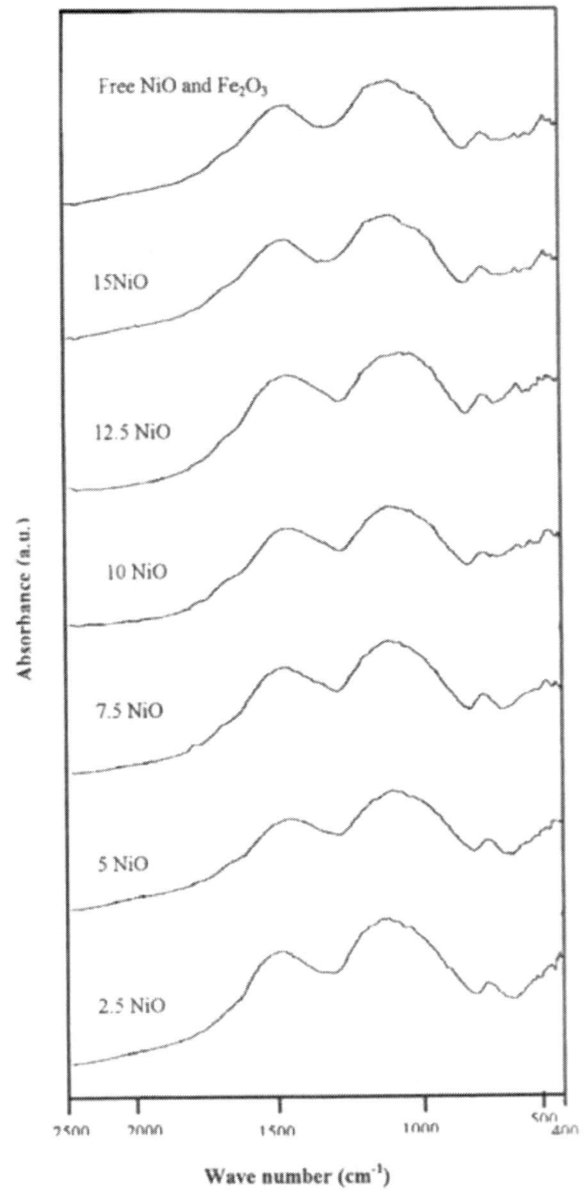

Figure 1. The IR spectra of the investigated glass samples.

The absorption band at 500 cm^{-1} shifted to higher frequencies of 530 and 580 cm^{-1} for the glass samples containing 5 and 7.5 mol% Fe_2O_3 respectively and then disappeared. The absorption band at 460 cm^{-1} shifted to 480 cm^{-1} for the sample containing 2.5 mol% Fe_2O_3 and remained constant until the10 mol% Fe_2O_3 sample before shifting back to 470 cm^{-1} in the glass sample containing 12.5 mol% Fe_2O_3.

The infrared spectra of borate glasses show three characteristic absorption bands at 700, 1260 and 1420 cm^{-1}. The absorption band at 700 cm^{-1} is due to

 vibrations, while the other two bands are due to

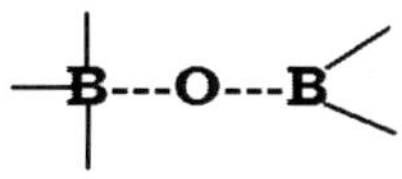 vibrations [12, 13].

The addition of Li_2O to the borate glasses gives extra oxygen atoms, which are accommodated into the network, by transferring some boron atoms from triangular BO_3 unite to tetrahedral BO_4 units. This could be deduced from the shift of the absorption band at 1420 cm^{-1} to a lower frequency. The vibrations of some boron non-bridging oxygen a groups in the form

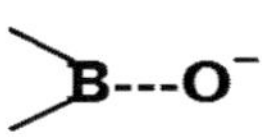 or 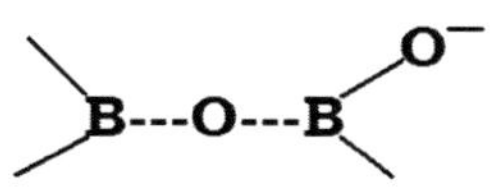 give a new band at 1010 cm^{-1} [12, 13].

In the infrared spectrum of the lithium borosilicate glass sample free from NiO and $Fe_2O_3$the absorption band at 1420 cm^{-1} may be assigned to B – O stretching vibration in two groups BO_3 and BO_4, also it may be assigned to 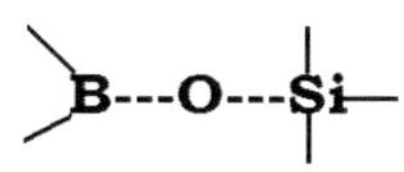 . The absorption band at 1030 cm^{-1} is due to stretching vibration of 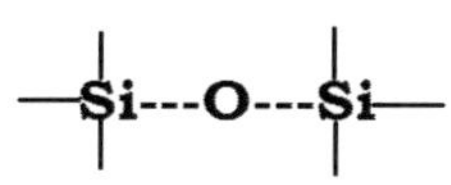 , also it may be attributed to the contribution of 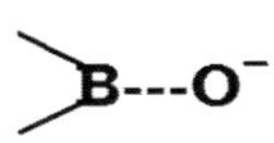 or 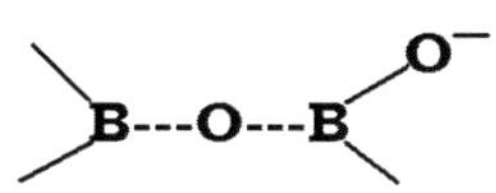 stretching vibrations [12, 13]. The bending vibration of 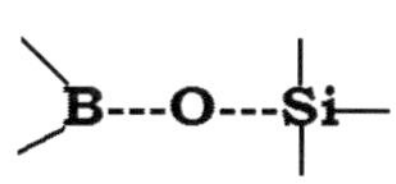 species appears at460 cm^{-1} [14-16]. The absorption band at700 cm^{-1} may be also due to B – O- B bond- bending vibration groups [17, 18]. The fact that all the mentioned bands are broad confirms the vitreous nature of the studied glass samples.

The band that appeared at 520 cm^{-1} is due to Si – O – Si vibrations (19), as well as the bands that appeared at 500 and 440 cm^{-1} which may also be attributed to the bending vibration of B – O bonds in a mode involving BO_4 units (17,18) and bending vibration of Si – O bonds.

When the nickel oxide was added on the expense of the lithium oxide it was observed that the band at 1420 cm^{-1} shifted to a lower wave number at 1400 cm^{-1} which is attributed to the change of BO_3 to BO_4 units [14-16].

When adding the iron oxide on the expense of the nickel oxide it can be observed that the band that appeared at 1020 $cm^{-1}$was shifted to a lower frequency with adding the iron oxide at the expense of nickel oxide up to 7.5 mol% Fe_2O_3, and then shifted to higher frequency in the glass samples containing more than7.5 mol% Fe_2O_3. This indicates that the number of non – bridging oxygen decreased up to 7.5 mol% Fe_2O_3, and then increased up to 12.5 mol% Fe_2O_3. The decrease of non – bridging oxygens can be attributed to the substitution of octahedral nickel ions by the largely tetrahedral ferric ions i.e iron ions adopt a network forming role. Anew band at 580 cm^{-1}, which appeared in the infrared absorption spectra of the samples containing 2.5, 5, 7.5 mol % Fe_2O_3, may be due to FeO_6 groups (20). Also the bands that appeared at 430 and 410 cm^{-1} in the spectra of the samples containing 15 mol% NiO and at 420 cm^{-1} in the sample containing 5 mol% Fe_2O_3 may be due to Si – O – S vibrations [11].

The structural changes associated with the nickel and iron addition have been analyzed on the basis of the lithium borosilicate glasses containing two absorption bands related to BO_3 and BO_2O- triangles and tetrahedral (BO_4 −) groups using spectral de-convolution into their Gaussian components.

From the relative peak areas of BO_3 and BO_2O- (A_3) and BO_4- (A_4) peaks, the values of N_4 calculated as $A_4/ (A_4 + A_3)$. The values of N_4 versus Fe_2O_3 content are shown in Figure 2. For all the investigated glass samples, the N_4 values are lower than one, showing the found of BO_3 units in these glass structures. It is well known that in borate glasses there is an isomerization process between three- and four-coordinated boron species as follows:

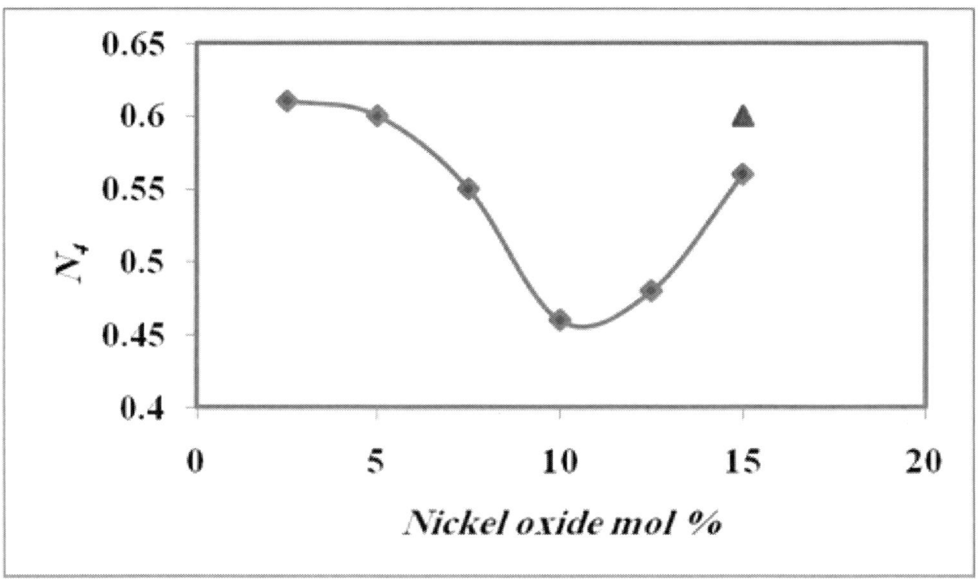

Figure 2. The values of N_4 versus NiO content. ▲ for sample free from Ni and Fe.

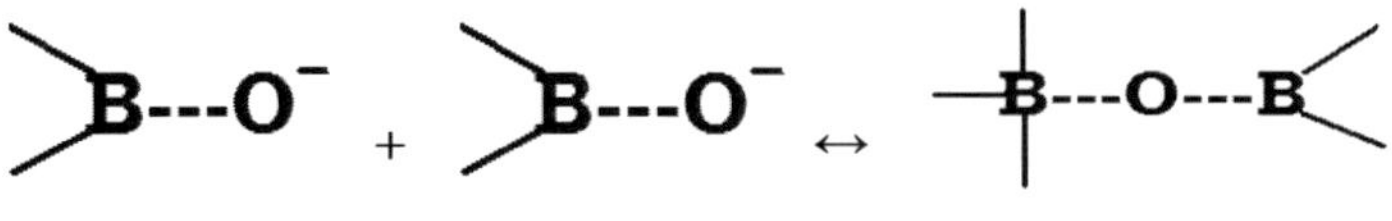

(1)

The value of N_4 in lithium borosilicate (free from nickel and iron) glass is $N_4 = 0.6$.

The N_4 values are decreased when NiO was introduced into the glass matrix (15 mol %, revealing a kind of equilibrium in the $BO_4/BO_4 + BO_3$ ratio and further modification in the glasses network, which means that BO_4 units increase. This may be because the oxide ions of Li_2O may be taken up by the NiO to make network forming nickel oxygen structure units, so gradually decreasing the total number of modifier ions and increasing the BO_4 in the networks units.

The N_4values decreased when the Fe_2O_3 is replaced NiO up to 7.5 mol % and then increased. The decrease of BO_3 can be attributed to the change of coordination of the added ions from the octahedral to the tetrahedral ferric state.

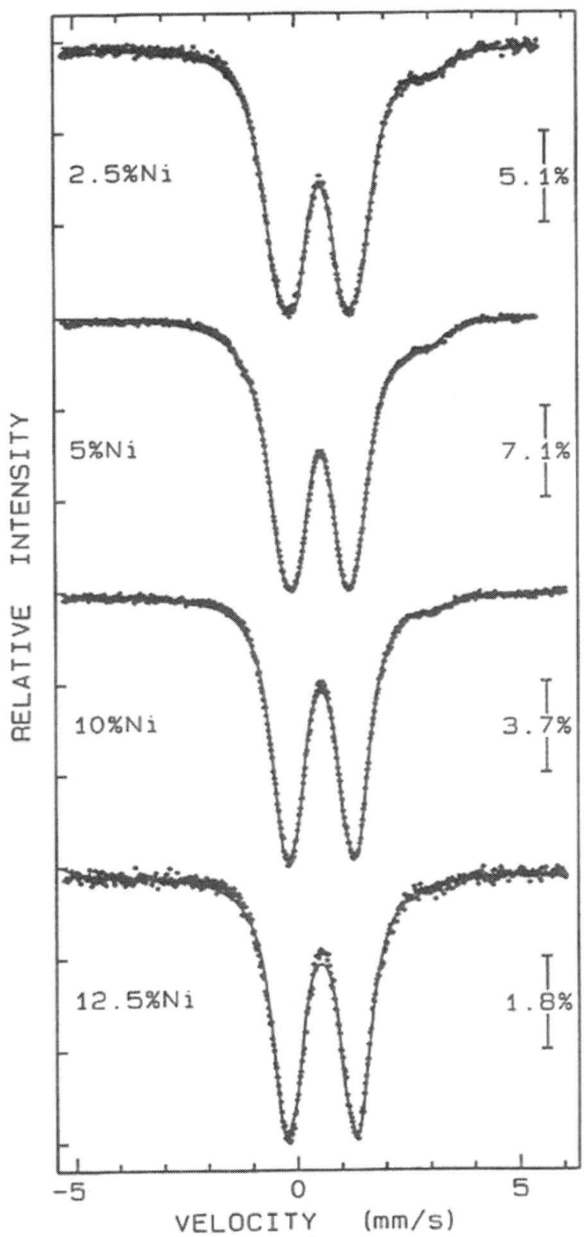

Figure 3. The Mössbauer spectra for the measured samples.

The Mössbauer spectra for the investigated glass samples are represented in Figure (3). The computer analysis indicated that the spectra of the samples containing 2.5, 5 and 7.5 mol% NiO are composed of two paramagnetic doublets, while the spectra of the samples which contain 10 and 12.5 mol% NiO are composed of only one single doublet. It was found that the ferric fraction $\{Fe^{3+}/\sum Fe\}$ increased with the gradual increase of nickel oxide up to 10 mol%, as shown in Figure (4a). The spectra of the samples containing 10 and 12.5 mol% NiO showed no ferrous ions.

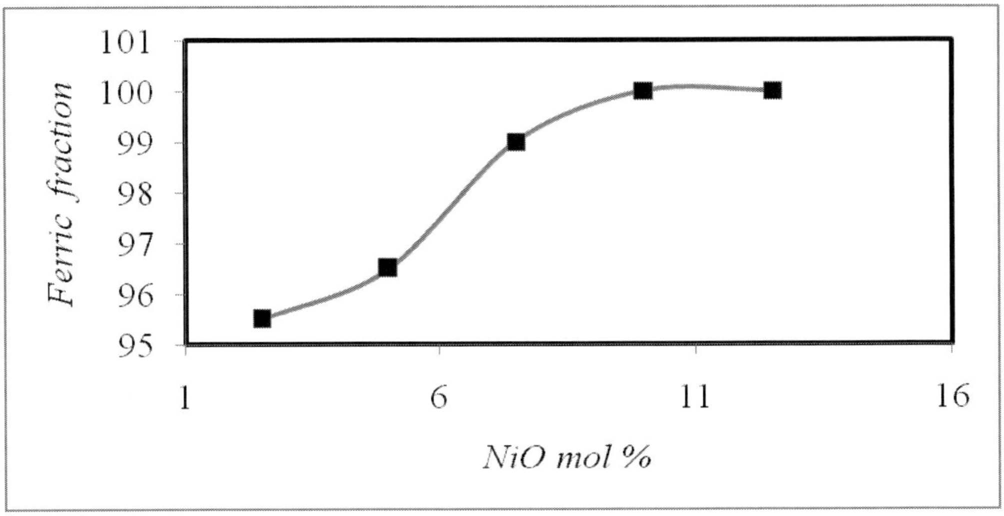

a

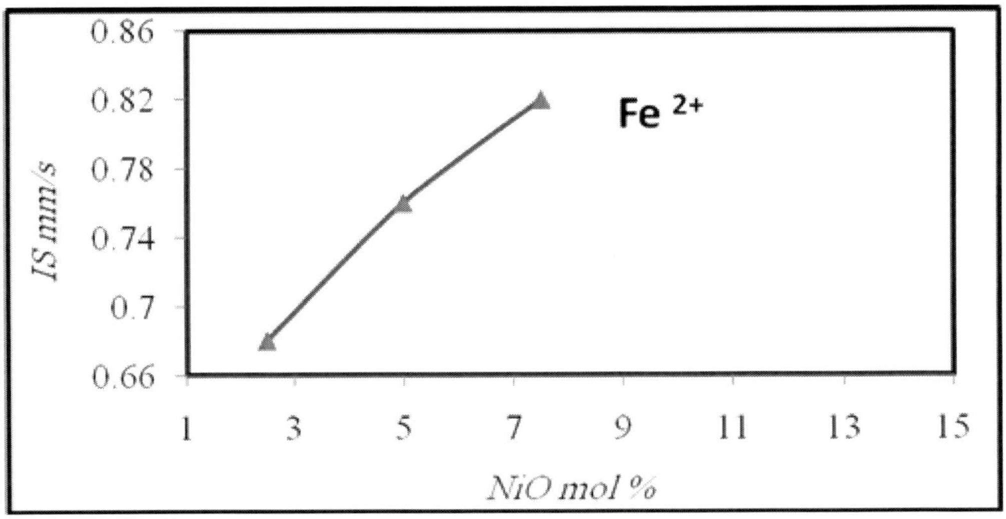

b

Figure 4. The Mössbauer parameters for the glass samples containing NiO mol% (a) the ferric fraction, (b) the isomer shift (δ).

The changes in the Mössbauer parameters as nickel oxide gradually increased are shown in Figure (4b). The isomer shift (δ) value for Fe^{3+} ions (as a major constituent) appears to increase gradually with the gradual increase of nickel oxide until it reaches a maximum at 7.5

mol% NiO, then it decrease up to 12.5 mol% NiO. The isomer shift value for Fe^{2+} ions (as a minor constituent) appears to increase gradually with gradual increase of nickel oxide up to 7.5 mol% NiO. The quadruple splitting value (Δ) of the major Fe^{3+}, and the minor Fe^{2+} ions showed an increase up to 7.5 mol% NiO, then (Δ) for the Fe^{3+} state decreases until NiO reaches 12.5 mol%, as shown in Figure (5a).

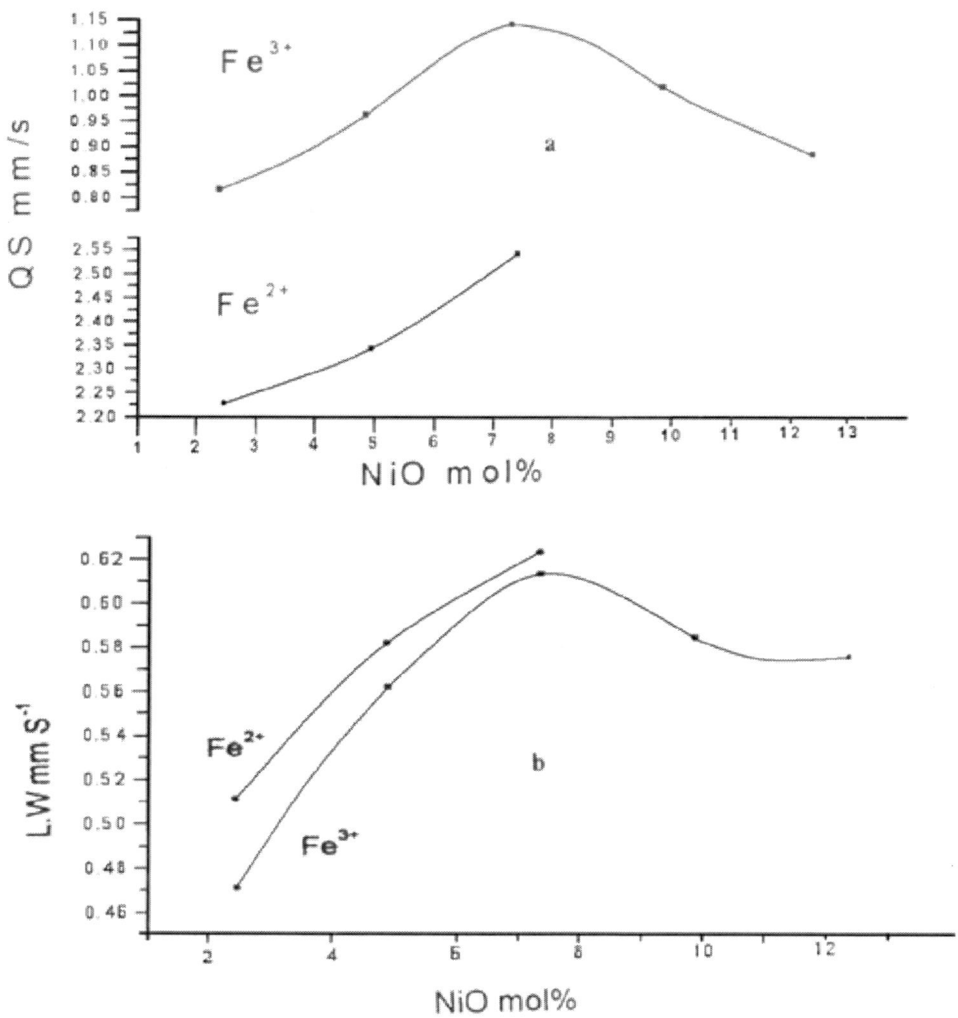

Figure 5. The Mössbauer parameters for the glass samples containing NiO mol% (a) the quadruple splitting (Δ) (b) the line width (Γ).

The line width (Γ) of both Fe^{2+} and Fe^{3+} increases with the gradual increase of NiO up to 7.5 mol%, as shown in Figure (5b). Above 7.5 mol% NiO, the line width (Γ) of the Fe^{3+} ions decreases with the gradual increase of NiO up to 12.5 mol%.

The obtained Mössbauer spectra for glass samples and their computer analysis indicated that the spectra of the measured samples containing 12.5, 10 and 7.5 mol% Fe_2O_3 are composed of two paramagnetic doublets, revealing that the iron ions occupy both ferrous and ferric states. Conversely the Mössbauer spectra of the samples containing 5 and 2.5 mol% Fe_2O_3 are composed of a single doublet, indicating that the iron ions occupy only ferric state. It can be concluded, from the values of the Mössbauer parameters, that all the ferrous cations are in the octahedral coordination state, while all the ferric ions are occupy the tetrahedral coordination state [20].

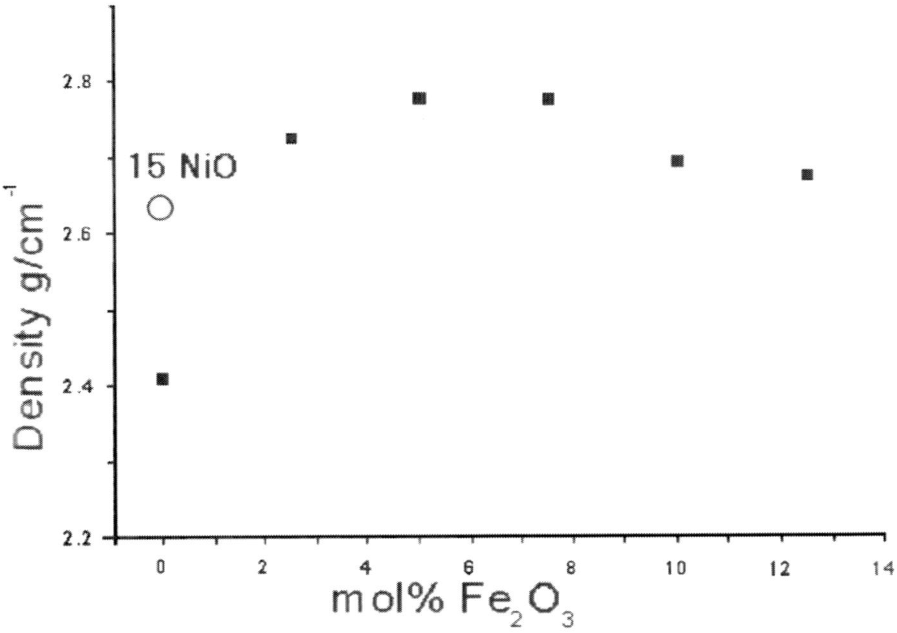

Figure 6. The density of glass samples containing Fe_2O_3 mol% Fe_2O_3.

The increase of the $\{Fe^{3+}/\sum Fe_{total}\}$ ratio with the gradual increase of nickel oxide up to 10 mol% NiO may be attributed to the decrease in the number of non – bridging oxygens which can be clearly observed from the increase of the values of the isomer shift [21].

The isomer shift values for Fe^{3+} ions increased with the gradual increase of nickel oxide up to 7.5 mol% NiO, and then decreased. The increase may be due to the decrease in the number of the non – bridging oxygen [22]. The isomer shift values for Fe^{2+} state also increased with the gradual increase of nickel oxide up to 7.5 mol% NiO, attributable to the decrease in the number of non – bridging oxygens [22].

The quadruple splitting value (Δ) of the major ions, Fe^{3+}, increased as NiO increased up to 7.5 mol% NiO and then decreased until the nickel oxide reached 12.5 mol% NiO, such an increase could be attributed to the decrease in the number of non – bridging oxygens [23]. The subsequent decrease could be due to the distortion of the oxygen polyhedra around the iron ions [22]. This may be due to the complete conversion of the octahedral Fe^{2+}ions to Fe^{3+} tetrahedral ions.

The line widths (Γ) of both Fe^{2+} and Fe^{3+} ions increase with the gradual increase of NiO up to 7.5 mol%. This increase indicates that nickel ions act to increase the asymmetry and the distortion of the glass network [24-26]. The line width of Fe^{3+} decreased with the increase NiO from 7.5 to 12.5mol%, indicating that the symmetry in this range is increased.

The density of a material is an important property, which relates directly to its structure and other properties. The density of the investigated glass samples is shown in Figure (6). The density of the glass sample free from nickel and iron oxides ($35Li_2O$, $31B_2O_3$, $34SiO_2$) was found to be 2.383 gm/cm^3, on adding nickel and iron oxides at the expense of lithium oxide the density increased to 2.67341 gm/cm^3. The increase in the density values may be due to the fact that atomic weights of nickel and iron atoms are greater than the atomic weight of lithium atoms. When introducing nickel oxide at the expense of iron oxide, the density slightly increased up to 7.5 mol% NiO, and reached a maximum value at 10 mol% NiO, and then decreased to 12.5 mol% NiO. The increase in the density values may be due to a decrease in the concentration of non – bridging oxygens, causing a decrease in the volume; this conclusion is in a good agreement with the FTIR and Mössbauer results. The decrease in the density values may be due to the transfer of some ions from the octahedral state to ions in the tetrahedral state, which causes an increase in the number of non – bridging oxygen, this causes an increase in the volume of the samples and a decrease in the densities as shown in Figure (6).

In previous work, I. Kashif et al. studied the structure and physical properties of Lithium borosilicate glasses containing Fe_2O_3. They found that the iron act as Fe^{2+} and Fe^{3+} in both tetrahedral and octahedral state. Also the Fe^{2+} / Fe^{3+} ratio equal 0.3 [27]. The drop of this ratio to 0.04 and change to tetrahedral only, it means that the Ni ions replacement the Fe^{2+} and enter the samples as Ni^{2+} in octahedral state.

CONCLUSION

According to the obtained results and the proposed discussion, it can be concluded that:

The addition of nickel oxide for lithium oxide decreases the number of non- bridging oxygens. The replacement of up to 7.5 mol% nickel by the same amount of iron decreases the non – bridging oxygen content which then increases again at higher iron levels. The N_4 values are decreased when NiO is introduced in the glass matrix, which means that the BO_4 units might increase. This may be because the oxide ions of Li_2O may be taken up by the NiO for network forming nickel oxygen structure units, which gradually decrease the modifier ions and increase the BO_4 in the networks units. The number of non – bridging oxygen decreased up to 7.5 mol% Fe_2O_3, and then increased up to 12.5 mol% Fe_2O_3. The decrease of non – bridging oxygen can be attributed to entering iron ions in the network former.

REFERENCES

[1] P.J. Bray, *J. Non-Cryst. Solids* 73 (1985) 19.

[2] W. Vogel, Struktur und Kristallisation der Gläser, VEB Deutscher Verlag fr Grundstoffndustrie, Leipzig, (1971) 98.

[3] S. Liu et al. *J. of Non-Cryst. Solids* 354 (2008) 956.

[4] M. S. Gaafar, S.Y. Marzouk, *Physica* B 388 (2007) 294.

[5] W. L. Konijnendijk, Ph.D. *Thesis, Philips Res. Suppl.* 1 (1975).

[6] Y. I. Jialiang, *J. Non-Cryst. Solids* 84 (1986) 114.

[7] K. El – Egili, *J. Physica* B 535 (2003) 340.

[8] A. S. Tenny, J. Wong, *J. Phys. Chem.* 56 (1972) 5516.
[9] W.A. Pliskin, H. R. Huffand, R. Burgess, *J. Electrochem. Soc.* 120 (1973) 506.
[10] A. A. Soliman, H. Farouk, I. Kashif, A. M. Sanad, *Czechoslovak J. of Phys.* 47 [5] (1997).
[11] S. A. Aly, A. A. Soliman, I. Kashif, A. M. Sanad, *Al-Azhar Bull. Sci.*8 [2] (1997) 493.
[12] H. Farouk, *Al-Azhar Bull. Sci.* 1 [2] (1990) 37.
[13] N. AElalaily, R. M. Mohamed *J. nucl. Mater.* 303 (2002) 44.
[14] H. S. Farhan, Ph.D. Thesis, *Al-Azhar Univ.*, Egypt (1988).
[15] I. Kashif, H. Farouk, A. M. Sanad , S. A. Aly and H. Farhan *Phys. Chem. Glasses* 32 [3] (1991) 87.
[16] N. AElalaily, R. M. Mohamed, *Mater. Sci. Eng.* B 98 (2003) 193.
[17] E. I. Kamitsos, A. P. Patsis, M. A. Kararkassides, G. D. Chryssikos, *J. Non-Cryst. Solids* 126 (1990) 52.
[18] K. El. Egili, *J, Physica B*, 325, 340-384(2003).
[19] Kashif, H. Farouk, A. A. Soliman, S. A. Aly, H. Zar El-Deen , A. M. Sanad, *J. Mat. Sci.* Eng. B 38 (1996) 217.
[20] R. Iordanova, Y. Dimitniev, V. Dimitrov & D. Klissurski, *J. Non- Cryst.Solids*, 167 (1994) 74.
[21] T. Nishida, Y. Takashima,(In industrial application of Mossbauer Effect) , New York (1986).
[22] Chul Sung Kim, Seung Wha Lee , Seung Lel Park, *J. Appl. Phys.* 79 (1996) 8.
[23] T. Nishida, T. Hirai,Y. Takashima, *Phys. Chem. Glasses* 22 (1981) 94.
[24] M. Sanad, I. Kashif, A.A. El- Sharkawy, A.A. El-Saghier, H. Farouk, *J. Mat.Sci.* 21 (1986) 3483.
[25] Kashif, Samy A. Rahman, A.G. Mostafa, E.M. Ibrahim, A.M. Sanad, *J. Alloys Compd* 450 (2008) 352.
[26] Kashif, S. M. Salem, A. A. Soliman, H. Farouk, A. G. Mostafa, S. A. Salah, A. M. Sanad, *J. Phys. Chem. Solids* 67 (2006) 2370.
[27] Kashif, S. Sh. Gomaa, A. G. Mostafa, S. M. Hamad, A. M. Sanad, *Phys. Chem. Glasses* 29, 2, (1988) 72.

In: Nanoscale-Arranged Systems for Nanotechnology
Editors: Kirill L. Levine
ISBN: 978-1-63482-353-1

Chapter 2

SURFACE AND TRANSPORT PROPERTIES OF THE PHASE CHANGE MATERIALS

O. S. Komarova**[,1], O. A. Martynova[1], A. V. Babichev[2,3], V. E. Gasumyants[1] and S. Vitusevich[4]

[1]St. Petersburg State Polytechnical University,
St Petersburg, Russia
[2]St. Petersburg Academic University, Nanotechnology Research and Education Centre,
RAS, St Petersburg, Russia
[3]Ioffe Physical Technical Institute, Russian Academy of Science,
St Petersburg, Russia
[4]Forschungszentrum Jülich, Peter Grünberg Institut, Jülich, Germany

ABSTRACT

In this work we have studied the $Ge_{0.15}Sb_{0.85}$ samples and transmission line module (TLM) structures of $Ge_2Sb_2Te_5$ compound grown on different substrates both in the amorphous and crystalline states. The resistance of TLM structures grown on the Si/SiO_2 was found to be less than ones grown on the Al_2O_3 substrate. The surface topology of the $Ge_{0.15}Sb_{0.85}$ samples of different thicknesses and phases was investigated. Besides, the resistivity and thermopower temperature dependences of $Ge_{0.15}Sb_{0.85}$ samples in the crystalline and amorphous states were measured and analyzed in order to clarify a type of the conduction process in samples of different phases.

INTRODUCTION

Phase change materials (PCMs) are materials that exist in at least two structurally different solid phases, namely, in the amorphous and crystalline states. Usually, the amorphous and crystalline phases demonstrate very different optical and electrical properties originated from a large difference in the crystalline structure. This can be used in

*E-mail:olik-komarik@rambler.ru.

technological applications for storing information, if it is possible to change repeatedly the state of a material between the two phases and both phases are stable at operating temperatures. Nonvolatile memory electrical devices based on PCMs cause now an increasing interest, in particular as a possible alternative to the standard silicon flash memory [1].

The first suggestions for using these materials for practical purposes have been made over 40 years ago [2], but interest to PCMs increased greatly in last years. This was due to the fact that the previously known PCMs were characterized by a rather slow crystallization speed. The situation changed dramatically after the discovery in 1987 of fast switching alloys based on the pseudobinary line between GeTe and Sb_2Te_3 [3]. As a result, from a practical point of view PCMs of such compositions as $Ge_2Sb_2Te_5$, $GeSb_2Te_4$, GeTe, Sb_2Te_3 and others are now the most promising materials for applications in the field of information storage.

PCMs have been successfully used to create rewritable DVD and Blu-ray discs [4], as well as a prototype of multi-layer memory cell based on the $Ge_2Sb_2Te_5$ compound [5]. However, there is no reproducible technology for creating structures based on the PCMs and characterized by a low noise and a high stability. The main reason for this situation is the lack of reliable and generally accepted data on the fundamental properties and charge carrier-system parameters in these materials. Electronic properties of PCMs are poorly investigated, although understanding of their features can be considered as a key for using these materials in solid-state memory devices. The data on the transport coefficients behavior in PCMs are fragmentary, moreover, some of these coefficients have not been investigated yet. Therefore, carrying out the systematic fundamental study of transport properties of different PCM-compounds is extremely important for the progress in the field of practical applications of these materials.

Samples and Experimental Details

In this work we have studied the $Ge_{0.15}Sb_{0.85}$ samples and transmission line module (TLM) structures of the $Ge_2Sb_2Te_5$ compounds grown on different substrates (Al_2O_3, Si/SiO_2) both in the amorphous and crystalline states. TLM structures have been made to investigate the contact properties and to perform the electrical characterization. All the investigated samples were prepared by magnetron evaporation. The thicknesses of the $Ge_2Sb_2Te_5$ deposited layer were 720 nm and 930 nm for the structures on the Al2O3 and Si/SiO_2 substrates, respectively. The distance between the contact pads of TLM structures decreases from 100 μm to 1 μm and their width decreases from 900 μm to 200 μm. Using the ion beam etching and lithography process allows us to obtain the high quality TLM structures. The rate of ion beam etching of $Ge_2Sb_2Te_5$ was determined to be 20.6 nm/min.

It is well known that electromigration of tellurium atoms to the contact area in $Ge_2Sb_2Te_5$ plays an important role and leads to a change in the properties of structures based on this compound. To minimize this effect, contacts were made from two layers (Ti 5 nm/Au 80 nm). However, one can choose another way, namely, using the binary $Ge_{0.15}Sb_{0.85}$ compound. This material also has a high switching speed (about 10 ns [6]) and, at the same layer thickness, a higher crystallization temperature compared to $Ge_2Sb_2Te_5$ (by $60^{o}C$). Note, that these are the $Ge_{0.15}Sb_{0.85}$ structures that were used to demonstrate a prototype of bridge memory cells with cross-section size as low as $20\times3nm^2$ [7].

$Ge_{0.15}Sb_{0.85}$ samples were grown in an amorphous phase on the Si/SiO_2 substrate and then crystallized by a slow heating in one case or by a fast heating (at a rate of about 1° C/sec.) in the second case to test the reproducibility of the results. The sample thicknesses were of 30, 60, and 600 nm. To investigate the thickness effect on the sample surface topology the AFM studies were carried out. Studying the samples topography was performed by the Dimension 3100 (Veeco) AFM in the semi-contact mode at room temperature using the RTESP probes with the curvature radius of 10 nm.

Measurements of the I-V characteristics of TLM structures were carried out using the HP 4156 Semiconductor Parameter Analyser directly on the TLM structure (using the probe station SussMicrotec). The resistivity temperature dependences on $Ge_{0.15}Sb_{0.85}$ samples were measured by the van der Pauw method [8] that allows us to solve the problem of arbitrary shape of samples. Determination of the thermopower coefficient, *S*, were carried out by the differential method relative to the copper electrodes.

RESULTS AND DISCUSSIONS

Measured I-V characteristics of TLM structures demonstrate a good scalability with the decrease of the length down to 1 μm for both crystalline and amorphous phases (see Figure 1 for the amorphous phase).

Note, that despite the different thickness of $Ge_2Sb_2Te_5$ (720 nm and 930 nm for the substrate of sapphire and Si/SiO_2, respectively) the final height of TLM mesas is the same, which allows us to compare quantitative the resistance of mesas with the same topological dimensions. It was found that the resistance of structures grown on the Si/SiO_2 substrate is less than the one of structures grown on the sapphire substrate. This could be due to a different adhesion of the contacts materials to $Ge_2Sb_2Te_5$ and stresses appearance that results in a defects formation leading to an activation of additional scattering mechanisms affected directly the resistance value of the structure.

To investigate the thickness effect on the sample surface topology the AFM studies were carried out. As an example, Figure 2 shows AFM images of the surface of the amorphous and crystalline $Ge_{0.15}Sb_{0.85}$ samples of 600 nm thickness. Surface profiles for 60 nm and 600 nm $Ge_{0.15}Sb_{0.85}$ samples in the amorphous and crystalline states are shown in Figure 3.

The surface of all the samples has a granular structure, particles sizes vary from 10 to 40 nm. Crystalline samples are characterized by a more uniform particle size than samples of the amorphous phase. At the same time, the surface of all the samples is quite smooth. In the crystalline phase the height variation is smaller than in amorphous samples for all investigated thicknesses (see Figure 3 as an example). For samples of the minimum thickness (30 nm) the surface profile is less smooth (the height dispersion is about 6-7 nm) indicating that the sample quality in this case is stronger affected by the substrate.

The transport properties (the temperature dependences of the resistivity and thermopower) of $Ge_{0.15}Sb_{0.85}$ samples with thicknesses of 60 nm and 600 nm were also measured in the temperature range of T = 77-300 K using the van der Pauw method [8]. The resistivity values were calculated taking into account a correction factor depended on the deference between the resistance values obtained by repeating measurements after switching polarities of both the current source and the voltage meter.

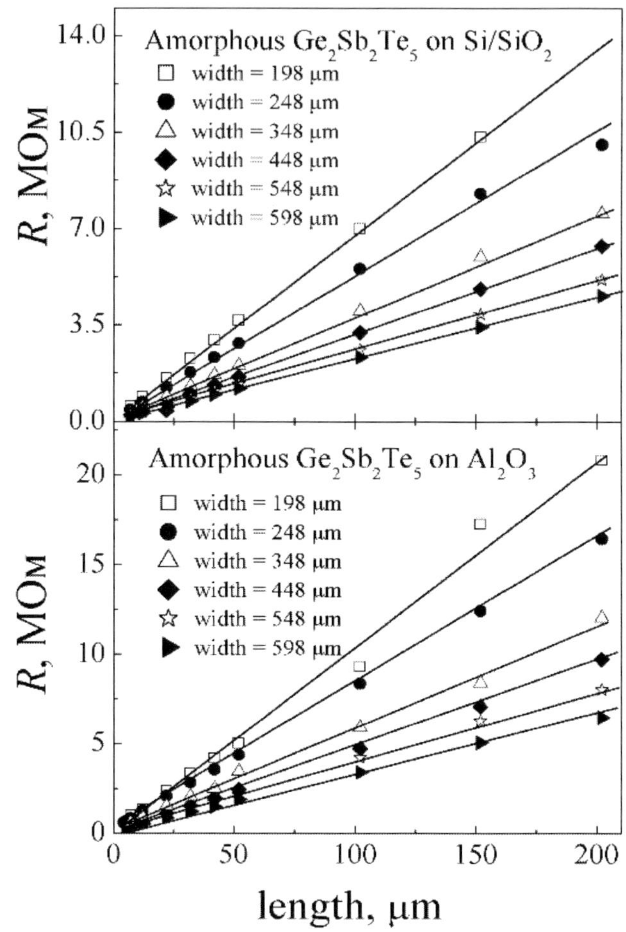

Figure 1. The resistance change depending on the distance between contacts in the TLM structures.

Figure 2. (Continued).

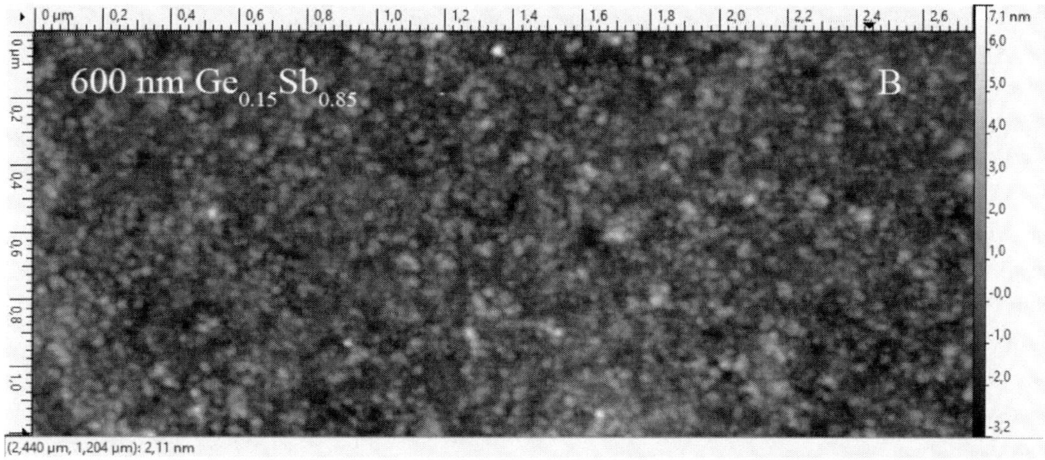

Figure 2. AFM images of the 600 nm $Ge_{0.15}Sb_{0.85}$ samples surface in an amorphous (A) and crystalline (B) states.

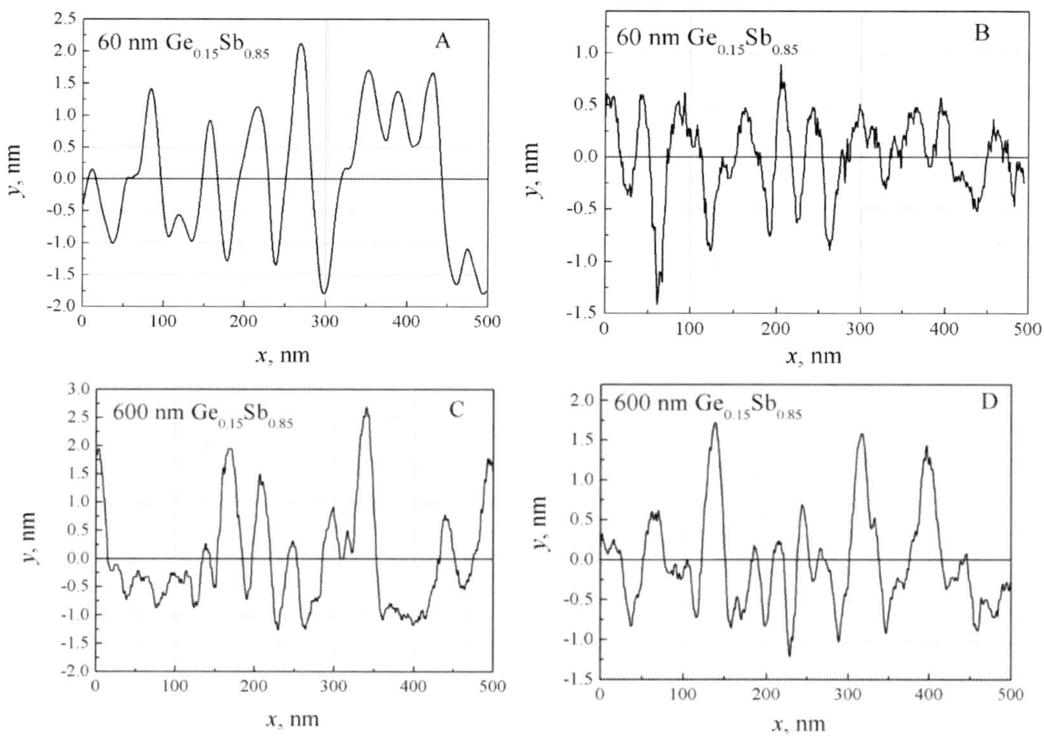

Figure 3. Surface profiles for 60 and 600 nm $Ge_{0.15}Sb_{0.85}$ samples in the amorphous (A, C) and crystalline (B, D) state.

The absolute resistivity values for amorphous samples were found to be by 3-4 orders of magnitude higher than for crystalline samples. Besides, the crystalline phase samples are characterized by nearly T-independent resistivity (see Figure 4 B for 60 nm thickness $Ge_{0.15}Sb_{0.85}$ sample), while in the amorphous state a strong rise of the resistivity with decreasing temperature was observed (see Figure 4 A for 600 nm thickness $Ge_{0.15}Sb_{0.85}$

sample) pointing to an activation type of the conduction process. To determine characteristics of this process the obtained temperature dependence was plotted in the semilogarithmic coordinates (see inset in Figure 4 A). One can clearly see that the $\rho(T)$ dependence in different temperature ranges is characterized by the two activation energies, E_a, namely, E_a = 0.13 eV at high temperatures and E_a = 0.007 eV at low temperatures.

The temperature dependences of the thermopower for amorphous and crystalline samples are shown in Figure 5. In the crystalline phase (see Figure 5 B for 60 nm thickness $Ge_{0.15}Sb_{0.85}$ sample) $S(T)$ is metal-like and the thermopower value at room temperature is about 5μV/K, that points to a degeneracy of the electron gas. In the amorphous phase we have observed a complex temperature dependence of the thermopower. In the temperature range of T = 150–300 K its values are positive, but at T < 125 K the S value becomes negative. This points to the existence of two type of carriers (electrons and holes).

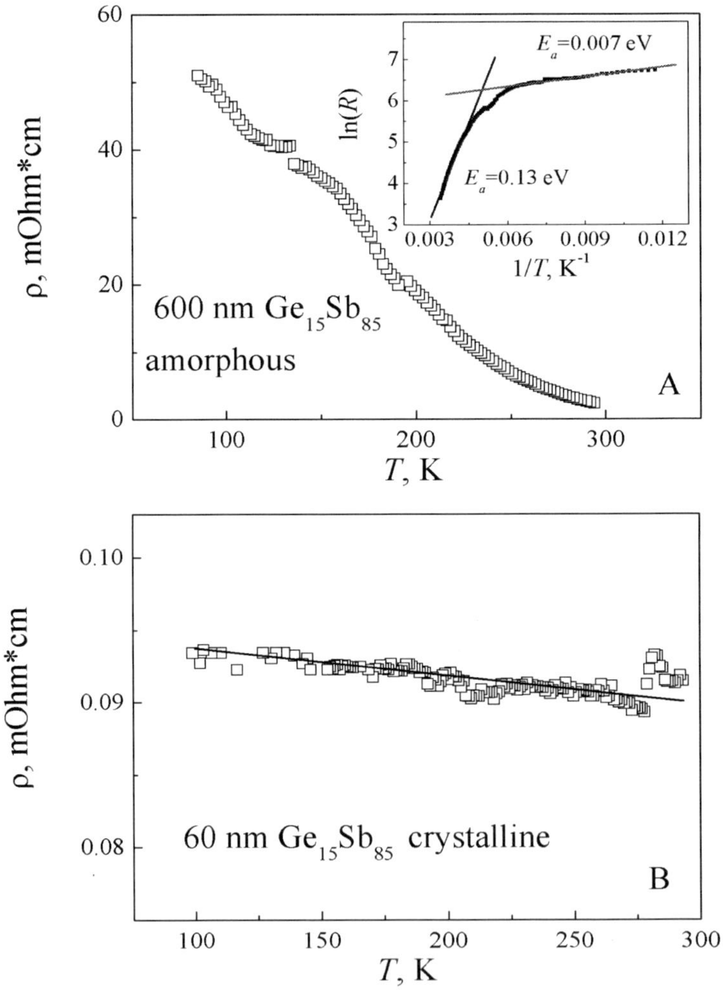

Figure 4. The temperature dependence of resistivity for 600 nm amorphous (A) and 60 nm crystalline (B) $Ge_{0.15}Sb_{0.85}$ samples. The inset shows calculated ln(R) values versus 1/T.

Note, that the temperature of a transition to negative thermopower values is closed to that of changing the ρ(*T*) slope. Thus, we can conclude that the conduction process in the amorphous state is of the electron type at low temperatures, but of the hole type at high temperatures.

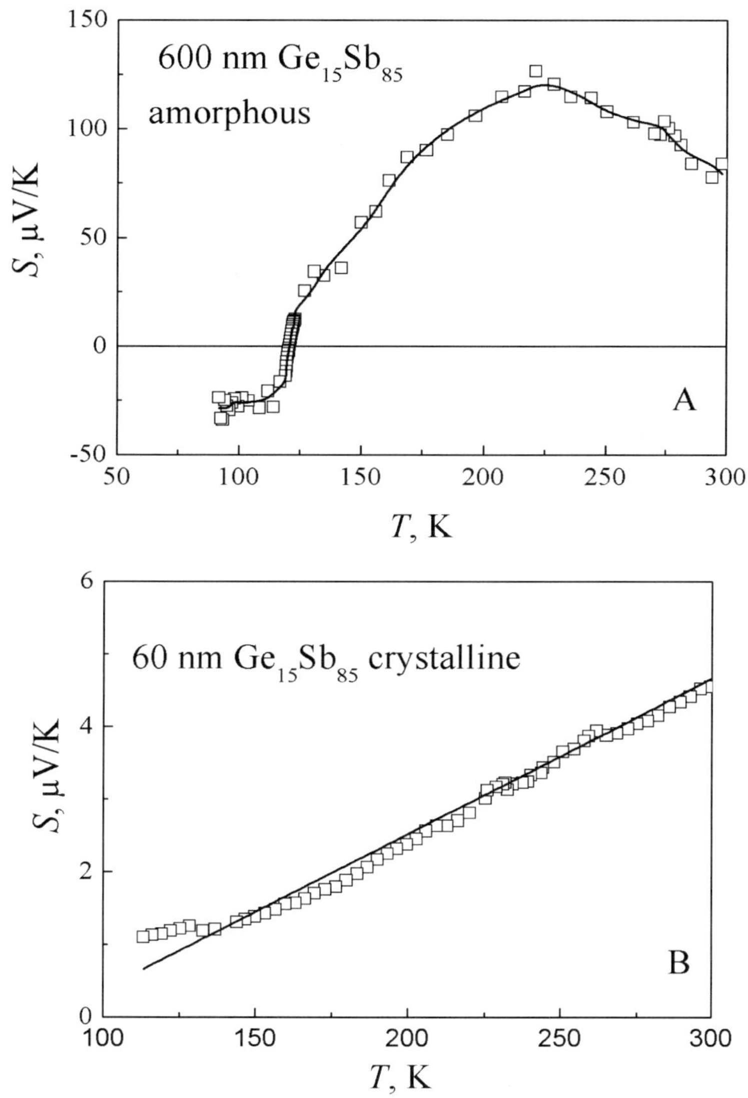

Figure 5. The temperature dependence of thermopower for 600 nm amorphous (A) and 60 nm crystalline (B) $Ge_{0.15}Sb_{0.85}$ samples.

CONCLUSION

In this paper, the following results were obtained:

1. Samples of $Ge_{0.15}Sb_{0.85}$ and $Ge_2Sb_2Te_5$ TLM structure in the amorphous and crystalline states on various substrates (Al_2O_3, Si/SiO_2) were prepared.
2. The measurements of I-V characteristics of TLM structures were carry out and the resistance of structures grown on the Si/SiO_2 substrate was found to be less than in case of using the Al_2O_3 substrate.
3. The surface studies of $Ge_{0.15}Sb_{0.85}$ samples of different thicknesses in both amorphous and crystalline phases were carry out. It was found that all the samples have a granular structure, but their surfaces are quite smooth, although for samples of 30 nm thickness the effect of the substrate on their quality is observed. Crystalline samples are characterized by a more uniform particle size than samples of the amorphous phase.
4. The transport properties of $Ge_{0.15}Sb_{0.85}$ samples in the crystalline and amorphous states were investigated. In the crystalline state the electron gas is degenerated, while in the amorphous state the conduction process has an activated character with the existence of different types of carriers at low and high temperatures.

ACKNOWLEDGMENTS

This work is supported by the grant of the President of Russian Federation for young scientists (No. MK-4608.2012.2). Authors thank Dr. P. Brunkov for performing the AFM measurements. This part of work was made on the equipment of the Joint Research Centre «Material science and characterization in advanced technology» (Ioffe Institute, St. Petersburg, Russia).

REFERENCES

[1] Raoux, S.; Wełnic, W.; Ielmini, D. *Chem. Rev. 2010, vol. 110, 240-267.*

[2] Ovshinsky, S. R. *Phys Rev Lett. 1968, vol. 22, 1450-1453.*

[3] Yamada, N.; Ohno, E.; Akahira, N. et al. *Jap. J. Appl. Phys. 1987, vol. 26-4, 61-66.*

[4] Wuttig, M; Yamada, N. *Nature mat. 2007, vol. 6, 824-832.*

[5] Chen, Y.C. *Phase Change Materials: Science and App.* 2008, *331-354.*

[6] Krebs, D.; Raoux, S.; Rettner C.T.et al. *J. Appl. Phys. 2009, vol. 106, 054308.*

[7] Chen, Y. C.; Lin, Y.; Chen, S-H. et al. *European Phase Change and Ovonics Symposium. 2008, 1-8.*

[8] van der Pauw, L. J. *Philips Res. Rep. 1958, vol. 13, 1-9.*

In: Nanoscale-Arranged Systems for Nanotechnology
Editors: Kirill L. Levine
ISBN: 978-1-63482-353-1

Chapter 3

Temperature Influence on the Rheology of Water-Based Drilling Fluids with Nano-Sized Polymer Additives

Mihaela Manea*
M-I SWACO, a Schlumberger Company
Schlumberger, Ploiesti West Park, Comuna Aricestii Rahtivani,
DN 72 Ploiesti Targoviste, Romania

Abstract

This paper gives an insight of water-based drilling fluids rheology when temperature is increased. The main goal is to approach formulations that can be used in drilling HTHP wells starting from using conventional polymers. The selected polymers for the study were carboxyl methyl cellulose, Xanthan gum, hydroxyl propyl guar and carboxyl methyl hydroxyl propyl guar. The guar gum derives were cross-linked.

The main novel aspect of this study is using the polymer additives at particles sizes of nano scale. These are breakthrough research ideas that can add more flexibility to drilling fluids formulations.

Introduction

The most recent and challenging issue in drilling fluids design is drilling HTHP wells. Considering the many difficult aspects of this matter, there are heavy conditions imposed to HTHP drilling fluids systems [1]. It is known that drilling fluids are dispersed systems that work at high depths in the borehole, meaning 7000-8500m underground, where there is a temperature gradient of 30°C/km. These facts expose the HTHP drilling fluids to temperatures up to 260°C and pressures of almost 30,000psi in the well.

There has been extensive research on drilling fluids that would perform in good parameters under HTHP conditions. Mostly, the drilling fluids created for this kind of

*E-mail: mmanea@slb.com.

environment are the non-aqueous fluids muds (NAF), i.e., RHADIANT, RHELIANT [2, 3]. In addition, there are also the water-based muds (WBM) that, despite of the water compressibility, can perform under HTHP conditions [4]. Those WBM systems are chosen mainly for environmental and economic reasons.

Regardless of the drilling fluid type, it is important to understand its chemistry and, also, to select the proper additives. For both cases, NAF and WBM, the main additives used are polymers.

Polymers are a class of organic compounds with special features, therefore it is crucial to work with those who exhibit thermic stability and maintain their unique properties under the HTHP conditions imposed.

The main focus is put on the drilling fluids rheology, namely on the aqueous polymer dispersion rheology.

In this study, only temperature effect was considered. The reason for this is that, first, rheology is greatly affected by temperature and, secondly, that, among all technological parameters, rheology is the most sensitive to any chemical addition.

The polymer rheology was been widely studied as far as the temperature effect [5, 6, 7], cross-linker presence [8, 9], pH changes [10] and so establishing the endurance to different treatments of various polymer dispersions.

The present study chose two classic drilling fluids additives, carboxyl methyl cellulose (CMC) and Xanthan gum, together with another two guar gums derives, hydroxyl propyl guar (HPG) and carboxyl methyl hydroxyl propyl guar (CMHPG). The last two polymers were studied both simple and cross-linked. The chemical reaction of a typical cross-linkage between guar gum and a borate cross-linker is given below, in Figure 1.

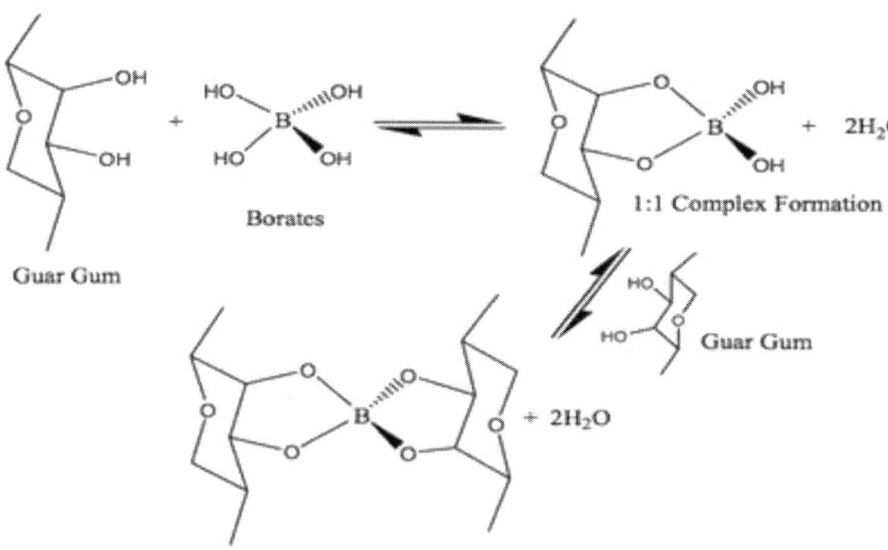

Figure 1. The equilibrium of borate ion complexation on a Guar Gum [11].

A major aspect of this study was to consider a nano size granulometry for the polymer additives mentioned. Applying nano technology to drilling fluids design became an interesting focus in the latest years [12, 13, and 14]. Primarily, the input of nano particles in drilling fluids consists in enhanced bit lubricity [15], decreasing sag and bridging flexibility.

The purpose of this study is to screen four polymer additives for their ability to form temperature resistant drilling fluids.

EXPERIMENTAL

The polymers used were ground with a Fritsch Pulverisette mill of agate bowl with balls and characterized by dynamic light scattering with a Malvern Zetasizer Nano ZS. The size of CMC was determined as an average diameter of 268nm, for the Xanthan gum the average particle size was 140nm, for HPG was 320nm and for CMHPG the average diameter was 119nm. The drilling fluids were prepared as aqueous dispersions adding the polymers under continuous mechanical stirring. The recipes contained the certain polymer studied plus a polymer fluid loss reducer [16]. The rheology was determined with a Brookfield PVS rheometer with the Rheovision software to analyze the data and a thermic bath controlled by a temperature regulator. In order to observe the thixotropic features of the drilling fluids, the shear stress measurements were performed both when increasing and when decreasing the shear rate.

RESULTS

The drilling fluids prepared with HPG without cross-linker were analyzed first. A 0.5%wt concentration of HPG was used and the flow curves were measured at four different temperatures. The last temperature step, 80°C, corresponds to a depth of 2800m.

The same measurements were performed with CMHPG without cross-linker and, furthermore, rheology was studied for aqueous dispersions with CMC and Xanthan gum.

To showcase better and to compare the flow performances of the soluble polymers used to prepare drilling fluids, their flow curves were represented in the same coordinates for each temperature. Figures 2, 3, 4 and 5 present those rheograms.

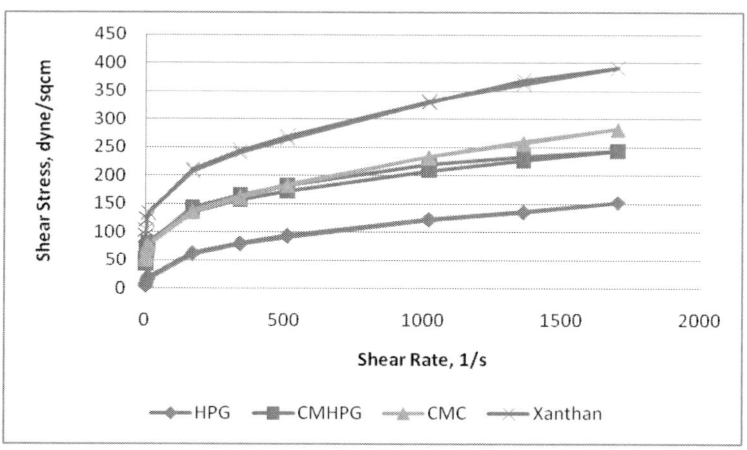

Figure 2. Rheograms at 20°C.

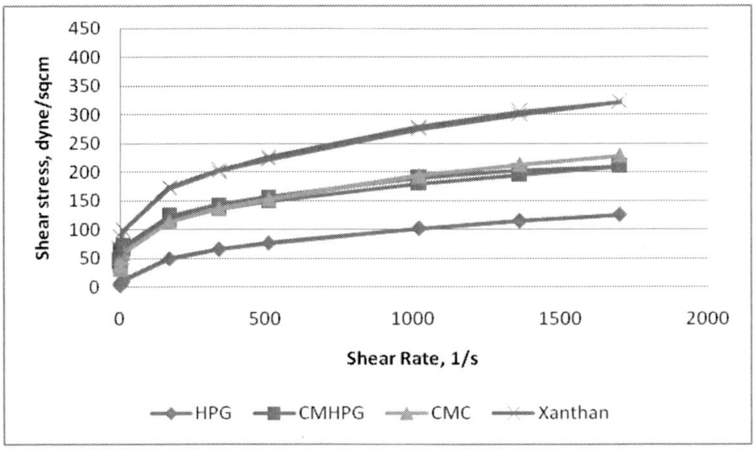

Figure 3. Rheograms at 40°C.

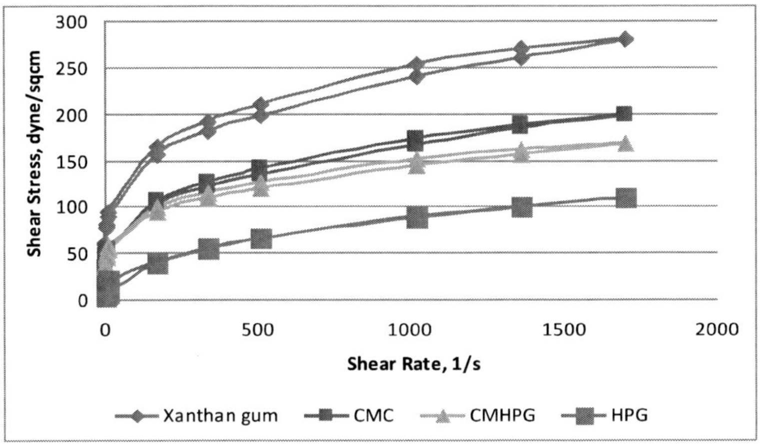

Figure 4. Rheograms at 60°C.

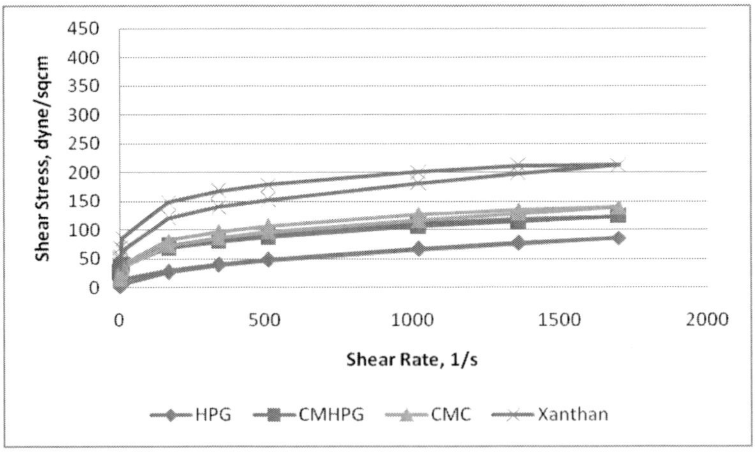

Figure 5. Rheograms at 80°C.

Comparing Figures 4 and 5, a slight hysteresis appears for the CMC and Xanthan gum prepared dispersions. This means an improvement in suspension capacity for these drilling fluids.

For the aqueous dispersion containing CMC, 1%wt, although temperature has the same effect of decreasing rheology, the values for shear stress maintain higher than in the case of guar gum derives, as presented in all figures above.

The common feature of all rheograms presented above is the order in which the shear stress values increase, namely fluids with HPG are the lowest, and the ones with CMHPG and CMC are very similar, while fluids with Xanthan gum have the highest values. To take into account that the mass concentrations for guar gum derives are 0.5%wt, while for the other two polymers were 1%wt.

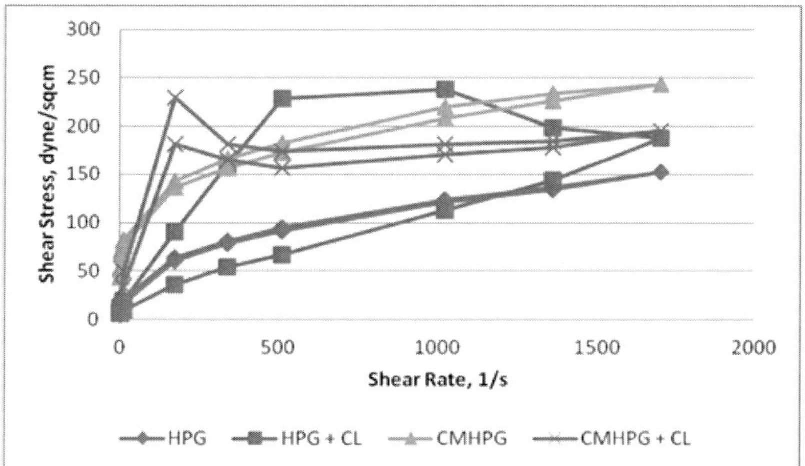

Figure 6. Rheograms at 20°C.

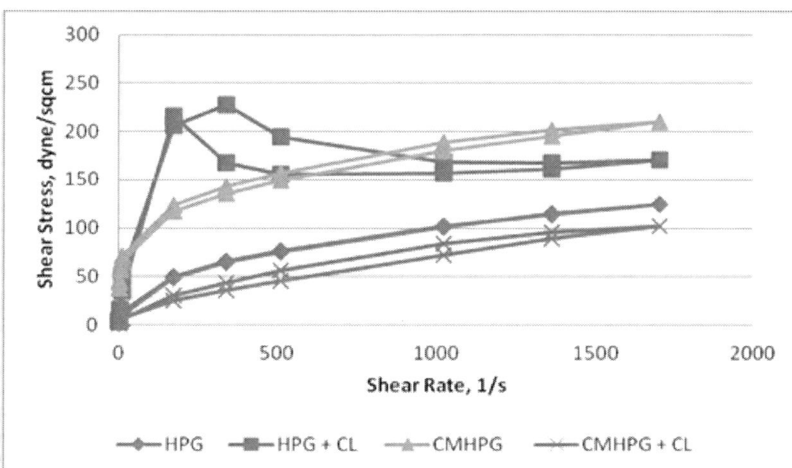

Figure 7. Rheograms at 40°C.

In the cross-linkage step of guar gum derives, the first experimental study focused on determining the appropriate polymer/cross-linker ratio for rheological purposes. For this, the soluble polymer concentration was brought down to 0.3%wt. The concentration of cross-linker was established at 0.3%wt. For a better appreciation of temperature and cross-linker presence influence on the water-based drilling fluids prepared with soluble polymers, derives of guar gum, the rheograms in Figures 6, 7, 8 and 9 are presented.

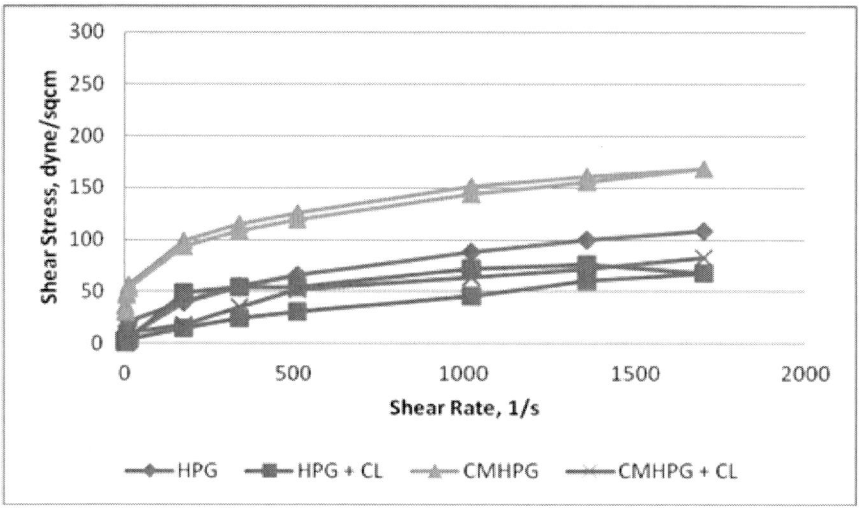

Figure 8. Rheograms at 60°C.

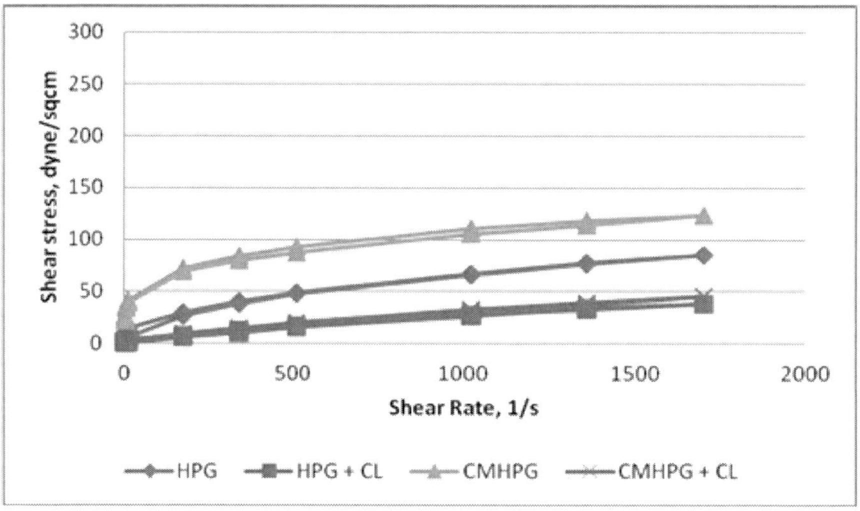

Figure 9. Rheograms at 80°C.

The flow curves have a fair constancy of flow parameters for the cross-linked fluids at moderate temperatures. At higher temperatures, the most resilient systems prove to be the dispersions prepared with non-cross-linked polymers.

For the cross-linked polymer dispersions, it can be seen that the flow properties are very good, although they progressively decrease with temperature. An important aspect is the appearance of a significant hysteresis, meaning thixotropic properties, that maintains itself up to 60°C. It is noticeable that none of these flow curves lose the curve profile which demonstrates a viscoelastic behavior; therefore, no system becomes unstable.

Conclusion

What is valid for all flow curves presented in figures that give the temperature influence on the drilling fluids prepared with soluble polymers is that in neither case the viscoelastic profile is lost. Therefore, at the temperatures analyzed, no dispersion loses stability, neither degrades thermic. Tough temperature has a negative influence on the rheology of all drilling fluids the Xanthan gum is acceptable at higher temperature values. This fact recommends it for usage at drilling fluids preparation when drilling deep intervals; the guar gum derives are recommended when drilling surface intervals.

Another argument to sustain the use of Xanthan gum aqueous dispersions at 80°C is that it is the only polymer that still has suspension capacity, namely hysteresis shown in the rheograms. If a cross-linker is added to the systems where guar gum derives are used, the rheological properties can be significantly enhanced. In the presence of a cross-linker, the polymer concentrations can be reduced a lot, keeping the flow parameters superior to those of the systems without cross-linker.

Acknowledgments

The author recognizes financial support from the POSDRU/89/1.5/S/54785 project: "Postdoctoral Program for Advanced Research in the Field of Nanomaterials".

References

[1] Bland, R; Mullen G; Gonzalez Y; Harvey F; Pless M. HPHT Drilling Fluid Challenges. *One Petro,* 2006, 24-35.

[2] http://www.slb.com/~/media/Files/miswaco/brochures/RHADIANT.pdf.

[3] https://www.slb.com/~/media/Files/miswaco/brochures/rheliant_ms06096.pdf.

[4] Dorman, J. An Ultimate Challenging Issue: Drilling Fluids for Ultra High Temperature Wells. *Challenges,* March 2010, 4-11.

[5] Al-Muntasheri, A; Hussein, IA; Nasr-El-Din, HA; Amin MB. Viscoelastic Properties of a High Temperature Cross-linked Water Shut-off Polymeric Gel. *Journal of Petroleum Science and Engineering*, 2007 Vol. 55, 56-66.

[6] Bradley, T D; Ball, A; Harding, SE; Mitchell, JR. Thermal Degradation of Guar Gum. *Carbohydrate Polymers,* 1989 Vol. 10, 205-214.

[7] Srichamroen, A. Influence of Temperature and Salt on Viscosity Property of Guar Gum. *Naresuan University Journal*, 2007 Vol. 15(2), 55-62.

[8] Morfopoulou, C I; Andreopoulou , AK; Daletou, MK; Neophytides, SG; Kallitsis, JK. Cross-linked high temperature polymer electrolytes through oxadiazole bond formation and their applications in HT PEM fuel cells. *J. Mater. Chem. A*, 2013, 1, 1613-1622.

[9] Pollino, JM; Nair, KP; Stubbs, L P; Adams J; Weck, M. Cross-linked and functionalized 'universal polymer backbones'via simple, rapid, and orthogonal multi-site self-assembly. *Tetrahedron,* 2004 Vol. 60, 7205–7215.

[10] Mu, B; Liu, P. Temperature and pH dual responsive crosslinked polymeric nanocapsules via surface-initiated atom transfer radical polymerization. *Reactive and Functional Polymers*, 2012 Vol 72(12), 983–989.

[11] Bahamdan, A. Hydrophobic guar gum derivatives prepared by controlled grafting processes for hydraulic fracturing applications, Louisiana State University, 2005.

[12] Amanullah, M; Al-Tahini, AM. Nano-technology – Its Significance in Smart Fluid Development for Oil and Gas Field Application. *Society of Petroleum Engineers (SPE),* 2009.

[13] Amanullah, M. Preliminary Tests Results of nano-based Drilling Fluids for Oil and Gas Field Applications. *Society of Petroleum Engineers (SPE),* 2011.

[14] De Stefano, G; Stamatakis, E; Riley, M; Young, S; Price, K; Friedheim, J; Quo, G; Ji, L. Wellbore stability in unconventional shale - The design of a nano-particle fluid. *Society of Petroleum Engineers (SPE),* 2012.

[15] Ray, TW; et.al. Drill Bit Having an Improved Seal and Lubrcation Method Using Same. US7013998 B2, 2006.

[16] Manea, M. Characterization of a Biodegradable Polymer Used as Additive to Prepare Drilling Fluids. *Revista de Chimie,* 2009 Vol. 60 (11), 1231-1234.

In: Nanoscale-Arranged Systems for Nanotechnology
Editors: Kirill L. Levine
ISBN: 978-1-63482-353-1

Chapter 4

PHASE AND STRUCTURE TRANSITIONS IN NANOPARTICLES OF SEMICONDUCTORS WITHIN POROUS DIELECTRIC MATRICES

S. Khanin[1], V. Solovyev[2], S. Trifonov[2] and V. Veisman[2]
[1]Herzen State Pedagogical University, St. Petersburg, Russia
[2]Pskov State University, Pskov, Russia

ABSTRACT

Semiconductor-to-metal phase transition in "vanadium dioxide–opal" ($VO_2 / opal$) and "vanadium dioxide – porous glass" ($VO_2 / glass$) nanocomposite materials and semiconductor-to-insulator structural transition in "iodine – AFI zeolite-like aluminophosphate" (I/AFI) at $T \approx 340\,\mathrm{K}$ have been compared. Experimental results demonstrate pronounced modification of optical and electrical properties of all the samples under study that is of great practical interest for the prospective possibility to design "smart nanocomposites" possessing bistability of physical characteristics.

In particular, one can use nonlinear optical limiting of electromagnetic radiation by vanadium dioxide nanoparticles in matrix composites for optical switching in information processing as well as for effective fast overload protection of different optical devices. On the other hand, posistor effect in I/AFI composite nanostructured materials makes it possible to construct temperature sensing devices for temperature measurement and control.

1. INTRODUCTION

In the last decades nanoporous dielectric matrices (e.g., opals, zeolites, porous glasses) have attracted considerable attention of physicists and chemists due to their applications for novel composite materials [1, 2, 3] and 3-dimensional photonic crystals [4, 5, 6]. Phase and structural transitions studies in these advanced materials are of great practical interest because of the prospective possibility to design "smart nanocomposites" possessing bistability of optical and electrical properties under definite physical conditions. At the same time many of these matrix nanocomposites physical characteristics and dominating mechanisms of observed phenomena are still a challenge.

In the present work we compare semiconductor-to-metal phase transition in "vanadium dioxide – opal" ($VO_2 / opal$) and "vanadium dioxide – porous glass" ($VO_2 / glass$) nanocomposite materials and semiconductor-to-insulator structural transition in "iodine – AFI zeolite-like aluminophosphates" (I/AFI). In all these cases one can observe pronounced modification of optical and electrical properties of the samples under study at $T \approx 340\,\mathrm{K}$.

Structure of porous dielectric matrices, as well as methods of sample preparation and investigation have been described in [2, 3, 6].

2. Semiconductor-to-Metal Phase Transitions in Vanadium Dioxide Nanoparticles in Opals and Porous Glasses

Vanadium dioxide (VO_2) possesses electronic semiconductor-to-metal phase transition dramatically changing its refraction index and absorption coefficient [7, 8]. These interesting optical properties of vanadium dioxide (especially those of VO_2 – based nanostructures) seem to be prospective for effective molding of the light flow. One can consider regular matrix photonic composite "vanadium dioxide – opal" ($VO_2 / opal$) as a good example.

Reflectance spectra of $VO_2 / opal$ (Figure 1) demonstrate Bragg diffraction of electromagnetic wave on periodic three-dimensional opal – based photonic crystal, possessing photonic band gap. The influence of semiconductor-to-metal phase transition in vanadium dioxide nanoparticles on intensity, width and position of the Bragg resonance is clearly seen. The last effect is also shown in Figure 2.

Modification of the $VO_2 / opal$ composite reflectance spectra due to semiconductor-to-metal phase transition (Figures 1 and 2) correlates not only with the behavior of the VO_2 thin film refraction index (Figure 3) but also with the electrical conductivity hysteresis loop of $VO_2 / opal$ nanocomposite (Figure 4).

Step-like shape of the last hysteresis loop (Figure 4) seems to be due to the step-like size distribution of the pore diameters in regular opal structure. On the other hand, similar electrical conductivity hysteresis loop of "vanadium dioxide – porous glass" ($VO_2 / glass$) nanocomposite possesses no step-like shape [6] probably due to continuous pore diameter size distribution near the middle value of about 10 nm in porous glass [9].

The largely extended control on the functionality of optical composites is based on the plasmon resonance effects as well as on the light-induced modification of the light [10, 11].

One more interesting result of semiconductor-to-metal phase transition in $VO_2 / glass$ nanocomposites is the manifestation of the nonlinear optical properties of these "smart materials", making laser radiation limitation possible [12]. This nonlinear optical limiting of radiation by $VO_2 / glass$ nanocomposites takes place for nano- and picosecond laser pulses in visible (530 nm) and near IR (1060 nm and 1540 nm) spectral regions (Figure 5).

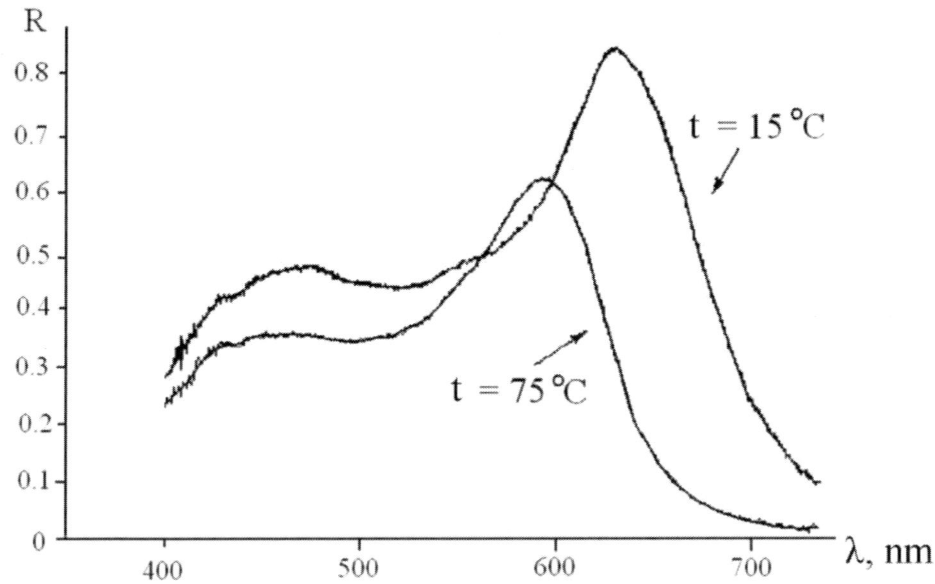

Figure 1. Reflectance spectra of $VO_2/opal$ nanocomposite before and after semiconductor-to-metal phase transition.

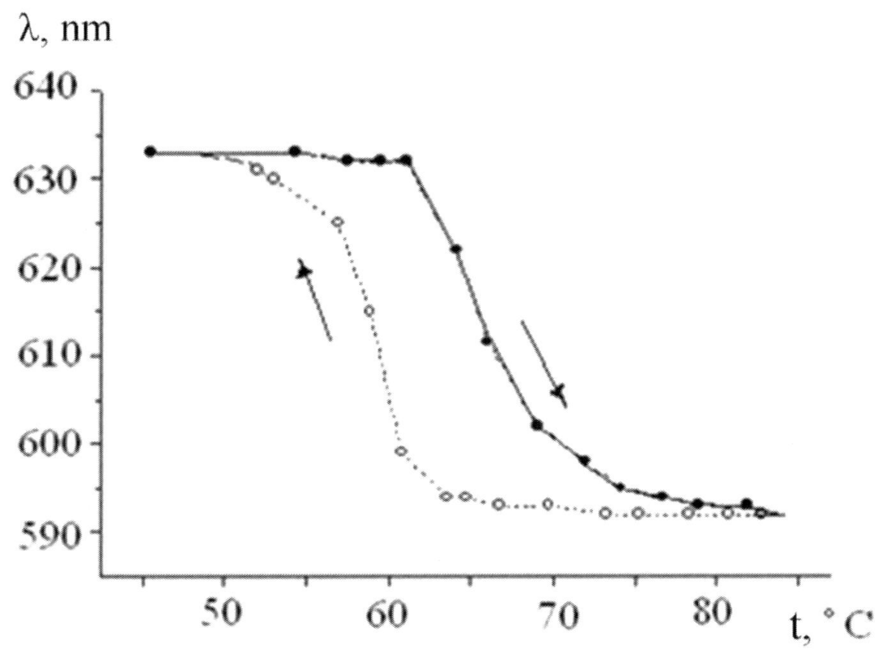

Figure 2. Hysteresis loop of Bragg resonance position in the reflectance spectra of $VO_2/opal$ photonic composite.

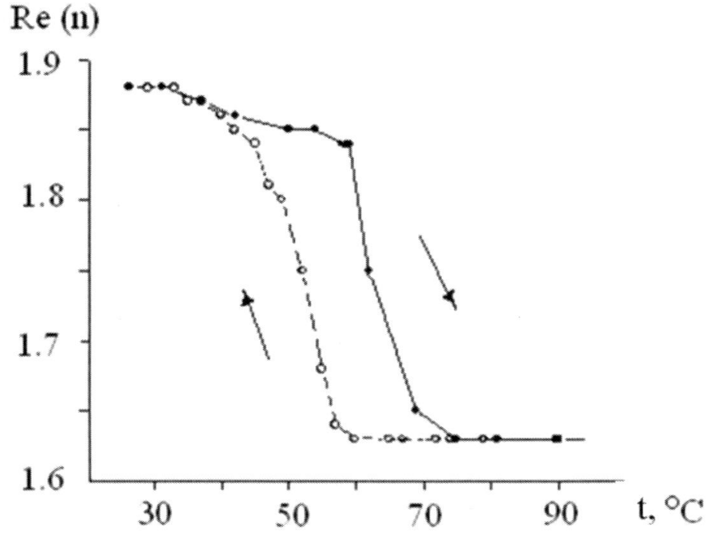

Figure 3. Hysteresis loop of the real part of refraction index of VO_2 thin film.

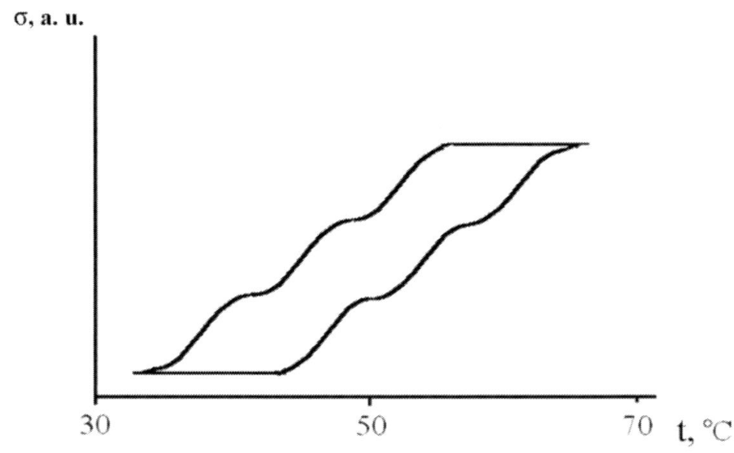

Figure 4. Electrical conductivity hysteresis loop of $VO_2 / opal$ nanocomposite.

The physical mechanism of this limitation is obvious: when nanostructured semiconductor turns into metal absorption coefficient dramatically increases. Nonlinear optical limiting of radiation by "vanadium dioxide – porous glass" nanocomposite is more pronounced for the wavelength $\lambda = 1540\,\text{nm}$ (Figure 5, *a*) mainly due to the existence of plasmon resonance in the near IR spectral region.

In conclusion, one can use nonlinear optical limiting of electromagnetic radiation by vanadium dioxide nanoparticles in matrix composites for optical switching in information processing as well as for effective fast overload protection of different optical devices.

Thus, semiconductor-to-metal phase transition occurs not only in bulk vanadium dioxide. It still takes place in vanadium dioxide nanoparticles with typical size $\sim 10\,\text{nm}$ both in opals and in porous glass matrices.

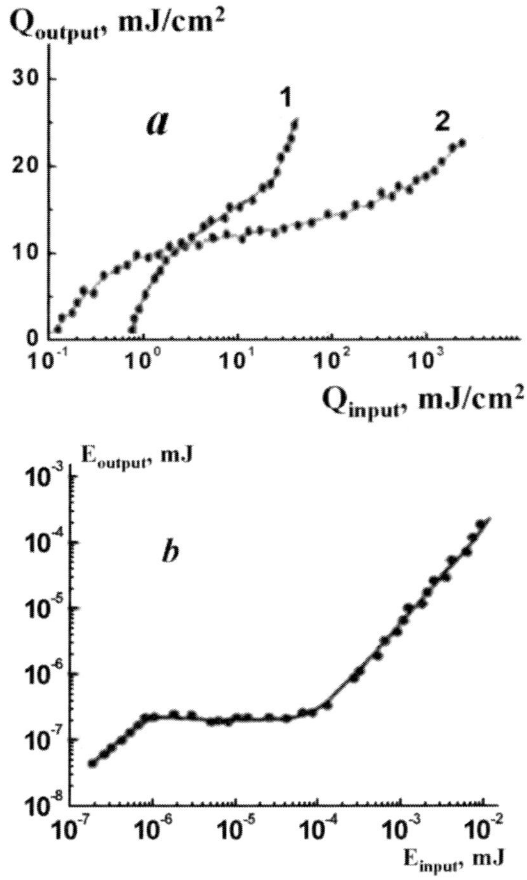

Figure 5. Nonlinear optical responses of $VO_2\,/\,glass$ optical composites irradiated by parallel light (*a*): 1) $\lambda = 530$ nm ; 2) $\lambda = 1060$ nm , and those under focused light irradiation (*b*), $\lambda = 1060$ nm , $\tau = 10$ ns .

In the next section we consider the situation when semiconductor-to-insulator structural transition arises due to another physical mechanism in the same temperature region after infiltration of porous matrix with smaller nanoparticles (typical size ~ 1 nm) of semiconductor, which possesses no similar transition in the bulk state.

3. Semiconductor-to-Insulator Structural Transition of Iodine Nanoparticles in AFI Zeolite-Like Aluminophosphates

Samples of I/AFI nanocomposite demonstrate large anisotropy of their optical absorption under polarized optical microscope in transmitted light with two perpendicular directions of incident polarization [13]. If electric vector $\vec{E}$ of electromagnetic wave is parallel to $\vec{c}$ - axis of AFI crystal, I/AFI single crystal is opaque, otherwise ($\vec{E} \perp \vec{c}$) it is transparent. This pronounced optical anisotropy is an important argument for the formation of quasi one-dimensional iodine chains inside AFI parallel channels. It should be mentioned that we did not find such optical anisotropy in I/A single crystals prepared in our laboratory by infiltration of three-dimensional system of voids and channels of zeolite type A nanoporous host matrix with iodine guest substance. Figure 6 demonstrates temperature dependence of the direct current $I_{||} = f\left(\frac{10^3}{T}\right)$ flowing through I/AFI composite along the $\vec{c}$ - axis of AFI crystal under constant applied voltage. The corresponding value of the activation energy $E = (0.7 \pm 0.1)$ eV calculated from Arrhenius plot in the temperature interval 310...340 K is also shown in the figure.

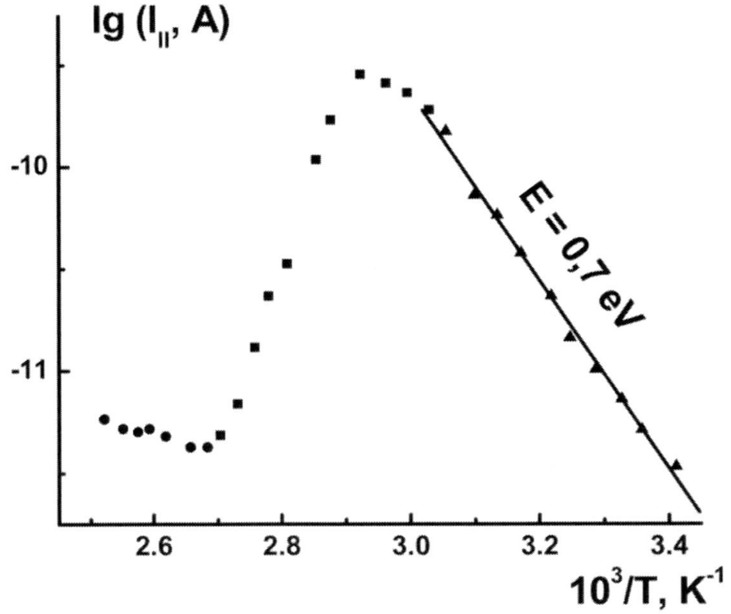

Figure 6. Arrhenius temperature dependence of the direct current flowing through I/AFI composite along the $\vec{c}$ - axis of the single crystal. Heating rate was about 1.5 K/min .

The most striking feature of this temperature plot is the pronounced peculiarity at $T_0 \approx 343$ K which could be observed when the temperature increases slowly and the current flows through I/AFI composite along the iodine chains.

We suppose that this peculiarity is due to the iodine structural transition from the chain structures to the molecular species, which was found recently at $\sim 70^\circ$C by means of optical characterization and phase dynamics of the iodine species in iodine@AFI composite system [14]. The possible mechanism of the charge transport in I/AFI composite at $T > T_0$ is the tunnelling of electrons between iodine molecular species.

Temperature dependences of the direct-current electrical conduction $G=f(T)$ and that of the electrical resistance $R=f(T)$ for I/AFI nanocomposite with low iodine concentration are plotted in Figures 7 and 8, respectively.

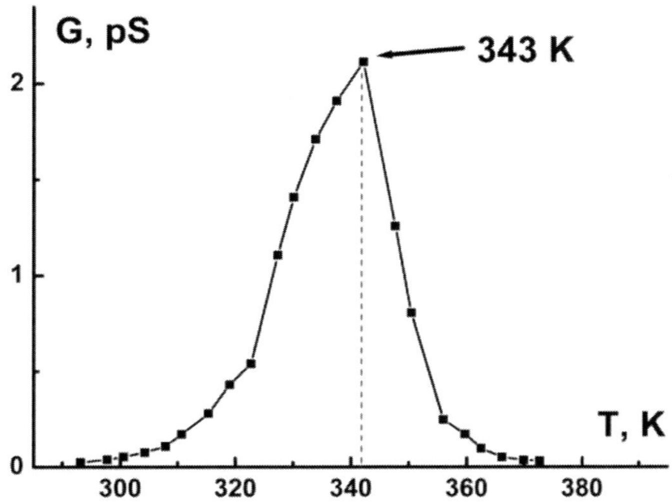

Figure 7. Temperature dependence of the direct-current electrical conduction $G=f(T)$ for I/AFI nanocomposite with low iodine concentration.

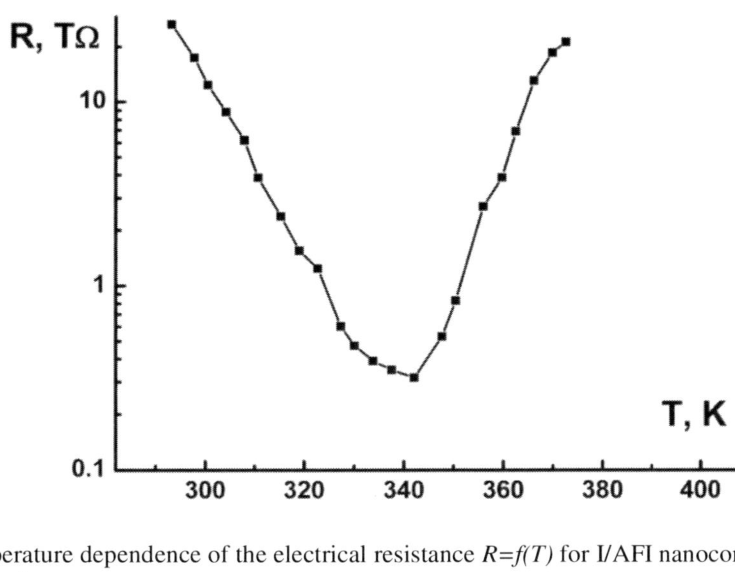

Figure 8. Temperature dependence of the electrical resistance $R=f(T)$ for I/AFI nanocomposite with low iodine concentration.

Thus, I/AFI nanocomposites demonstrate pronounced posistor effect in the temperature interval $343...373\ \mathrm{K}$, electrical resistance increasing with temperature by 10…100 times, temperature resistance coefficient being of about $+0.15\ \mathrm{K}^{-1}$. In principle, one can use this posistor effect in these "smart materials" to construct temperature sensing devices for temperature measurement and control.

It should be noted that bulk crystalline iodine (semiconductor with the band gap $E_g \approx 1.25$ eV) does not demonstrate any similar structural transition at $T_0 \approx 343$ K. So we can interpret phenomenon under study as the semiconductor-to-insulator structural transition in iodine nanoparticles in AFI zeolite-like porous matrix.

According to our experimental results the melting of iodine chain (aligned with the $\vec{c}$ - axis of the AFI single crystal) to molecular iodine has no noticeable effect on the direct current which flows across the $\vec{c}$ - axis (probably, also due to the tunnelling process). Thus, I/AFI single crystals possess pronounced anisotropy of their electrical properties. Moreover, this structural transition does not occur also during quick cooling of the "hot" I/AFI nanocomposite, infilled with disordered iodine molecular phase. After few days the ordered iodine molecular wires in I/AFI nanocomposite become partially reconstructed, and the phase transition at $\sim 70^{\circ}$C during heating could be observed again.

Conclusion

According to our experimental results bistability of physical properties of matrix nanocomposites containing nanoparticles of semiconductors may occur in two situations. Firstly, when phase transition is observed in bulk semiconductor, it may still take place in nanoparticles of this substance with typical size ~ 10 nm in porous matrices. Secondly, new structural transition may arise at reduction of the characteristic size of particles to ~ 1 nm even if semiconductor possesses no similar transition in the bulk state.

This work was supported by the Ministry of Education and Science of Russian Federation according to the programs "Development of Scientific Potential of Higher Educational Institutions" and "Scientific and Pedagogical Staff of Innovative Russia".

References

[1] Bogomolov, V. N. *Sov. Phys. Usp*. 1978, 21, 77– 83.

[2] Kumzerov, Y.; Vakhrushev, S. In *Encyclopedia of Nanoscience and Nanotechnology*; Nalwa, H. S.; Ed.; New York, US, 2004; Vol. 7, pp 811–849.

[3] Solovyev, V. G.; Ivanova, M. S.; Pan'kova, S. V.; Trifonov, S. V.; Veisman, V. L. In *Handbook of Zeolites: Structure, Properties and Applications*; Wong, T.W.; Ed.; New York, US, 2009; pp 77-99.

[4] Astratov, V. N.; Bogomolov, V. N.; Kaplyanskii, A. A.; Prokofiev, A. V.; Samoilovich, L. A.; Samoilovich, S. M.; Vlasov, Yu. A. *Il Nuovo Cimento* 1995, 17D, 1349 – 1354.

[5] Romanov, S. G.; Gaponik, N.; Eychmüller, A.; Rogach, A. L.; Solovyev, V. G.; Chigrin, D. N.; Sotomayor Torres, C. M. In *Photonic crystals: Advances in design, fabrication, and characterization*; Busch, K.; Lölkes, S.; Wehrspohn, R. B.; Föll, H.; Eds.; Weinheim, DE, 2004.

[6] Solovyev, V.; Kumzerov, Y.; Khanin, S. Physics of regular matrix composites (Electrical and optical phenomena in nanocomposite materials based on porous dielectric matrices); Saarbrücken, DE, 2011 (in Russian).
[7] Mott, N. F. Metal – Insulator Transitions; London, UK, 1974.
[8] Bruckner, W.; Opperman, H.; Reichelt, W.; Terukov, E. I.; Tschudnovskii, F. A.; Wolf, E. Vanadiumdioxide; Berlin, DE, 1983.
[9] Mikheeva, O. P.; Sidorov, A. I. *Zhurnal tekhnich. Fiziki* 2004, 74, 77-82.
[10] Neeves, A. E.; Birnboim, M. H. J. *Opt. Soc. Am*. B. 1989, 6, 787-796.
[11] Romanov, S. G.; Korovin, A.; Regensburger, A.; Peschel U. *Adv. Mater*. 2011, 23, 2515-2533.
[12] Vinogradova, O. P.; Obyknovennaya, I. E.; Sidorov, A. I.; Klimov, V. A.; Shadrin, E. B.; Khanin, S. D.; Khrushcheva, T. A. *Fizika Tverdogo Tela* 2008, 50, 734-740.
[13] Trifonov, S. V.; Vanin, A. I.; Veisman, V. L.; Gango, S. E.; Kondratyeva, M. N.; Solovyev, V. G. *Nanotechnics* 2011, 26, 78-82 (in Russian).
[14] Ye, J. T.; Tang, Z. K.; Siu, G. G. *Appl. Phys. Lett*. 2006, 88, 073114 (1)-073114 (3).

In: Nanoscale-Arranged Systems for Nanotechnology
Editors: Kirill L. Levine

ISBN: 978-1-63482-353-1

Chapter 5

METAL OXIDE SnO_2 - ZnO - SiO_2 FILMS PREPARED BY SOL-GEL

E. V. Abrashova**, *I. E. Kononova and V. A. Moshnikov
St. Petersburg State Electrotechnical University "LETI",
St. Petersburg, Russia

ABSTRACT

In this paper we investigate the features of formation of a hierarchical structure based on metal oxides $SnO_2 - ZnO - SiO_2$. The research shows, that pore size can be controlled by variation of the relative proportion of ZnO in solution. This fact is confirmed by calculating the fractal dimension of structures.

Keywords: metal oxides, fractals, sol-gel

INTRODUCTION

Multicomponent metal oxide materials obtained by sol-gel find many applications for gas sensitive layers [1-3], transparent conductors [4-6], multiferroics [7], thermo elements, catalysts [8,9] and others [10-12]. Sol-gel is one of the promising methods for obtaining such materials. The main advantages of coatings produced by sol-gel can be defined as relative chemical and morphological uniformity, the ability to precisely calculate exact composition of films, high processability and low cost [13-15].

From a molecular point of view the formation of nanocomposites, which occurs during the simultaneous reaction of hydrolysis and polycondensation can be considered as the nucleation and growth in the form of inorganic polymer. The aim of this work was to study the formation and growth of inorganic hierarchical porous structure system MeO – SnO_2 – SiO_2 [16, 17]. As MeO was chosen ZnO. Such structures obtained by sol-gel form branched

*Corresponding author: E V Abrashova, Postgraduate, St. Petersburg State Electrotechnical University "LETI", Prof. Popova st., 5, St. Petersburg, 197376, Russia. E-mail: katabr@mail.ru.

pore system, the nature and size of which are largely determined by technological conditions of production. Often impossible to obtain reproducible results strictly due to minor fluctuations of an external factors at different stages of formation. Therefore, it remains an open question about managing morphology introducing impurities in some way influencing the formation of cluster aggregates and their subsequent interaction.

EXPERIMENT

This work investigated the porous structure based on zinc oxide, tin and silicon. The experiment was considered by simultaneous maturation of 3 sets of solutions: based on tin dioxide, zinc oxide and based on total solution of tin dioxide and zinc oxide in the ratios 2:1, 1:1. 1:2. As the C_4H_9OH solvent was used, to improve stability and improve the adhesive properties to solution was added a small amount tetraetoxisilane (TEOS). TEOS was chosen to obtain the nanostructured films of silicon dioxide. TEOS solutions helped to improve the film-forming qualities and have the ability to spread over the surface of glass. The obtained sols were exposed to the normal conditions of exposure of 150 to 1500 hours. The solutions with different precursor concentrations were prepared and cast onto glass and silicon substrates. The resulting layers were spin-coated (3000 rpm) during 15 s and then annealed at different temperatures in the rang 100°C - 600°C. The surface morphology of thin-film nanomaterials was studied by tapping mode atomic force microscopy (AFM) using scanning probe nanolaboratory NTegra Therma (NT-MDT, Russia), study of the composition was performed by reflection hight-energy electron diffraction (RHEED) using EMR-102 (SELMI). Transmission spectrum of the samples was obtained using SF-56 spectrophotometer. As control samples for films deposited on glass and quarz were selected substrate without the film, i.e., uncoated glass and quartz glass, respectively, in order to avoid introducing systematic error due to influence of the substrate material.

RESULTS AND DISCUSSION

AFM analysis has shown that a solution based on tin dioxide matured , as expected, in accordance with the model of cluster-cluster aggregation, forming a labyrinth pattern (Figure 1a), then going to the percolation mesh (Figure 1b).

Studying the surface of samples obtaining from solutions of zinc oxide by AFM has shown that porous structure formation is not clear, but microcrystalline inclusions were detected whose density is increased by increasing the proportion of zinc oxide in solution.

While research by RHEED has detected that reflexes related to the zinc oxide phase clearly traced (Figure 2, 3a), from which we can conclude that ZnO micro-crystallites were formed. That is why we expected that ZnO micro crystallites will be formed in of tin dioxide and silicon dioxide matrix but studies by RHEED revealed that ZnO micro crystallites were not found (Figure 3 b).

Study of surface morphology for different relative proportions of tin dioxide and zinc oxide by AFM showed, that increasing the relative proportion of zinc oxide involves obtaining more rarefied porous structures. To quantify the porosity parameters were evaluated by various method fractal dimension using the Gwyddion (Table 1).

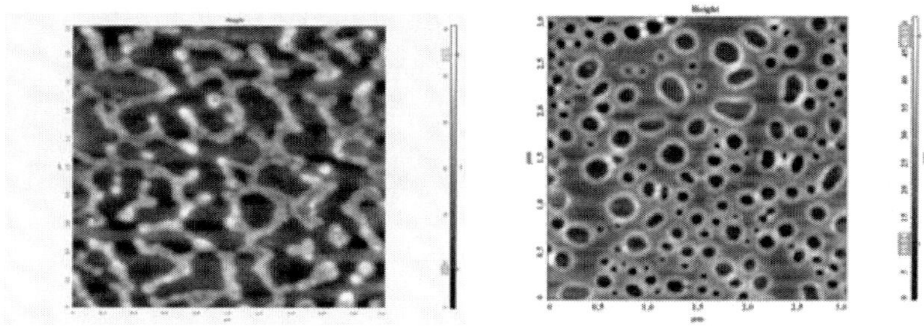

a) b)

Figure 1. Images of surface topography of the thin film of the solution composition on the basis of SnO_2 - 90%; SiO_2 - 10% by AFM: a) form a labyrinth structure, and b) form a percolation network structure.

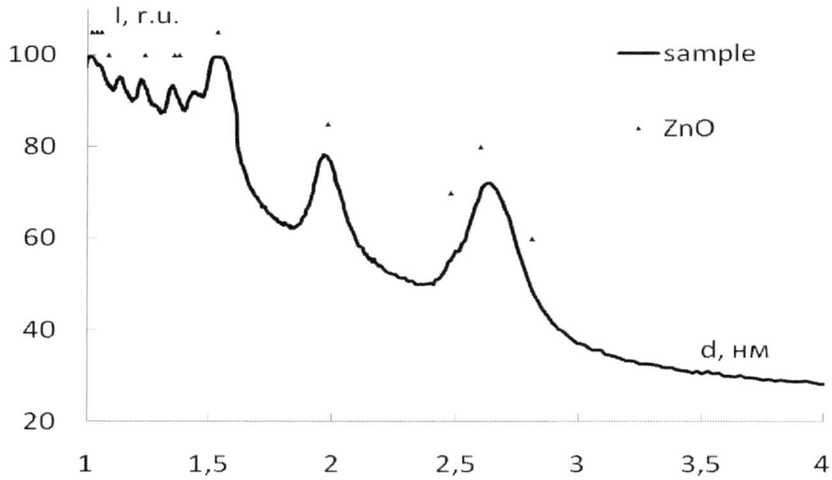

Figure 2. The intensity of reflections obtained by RHEED from the sample in relative units in comparison with the position of the peaks for the reference substance ZnO.

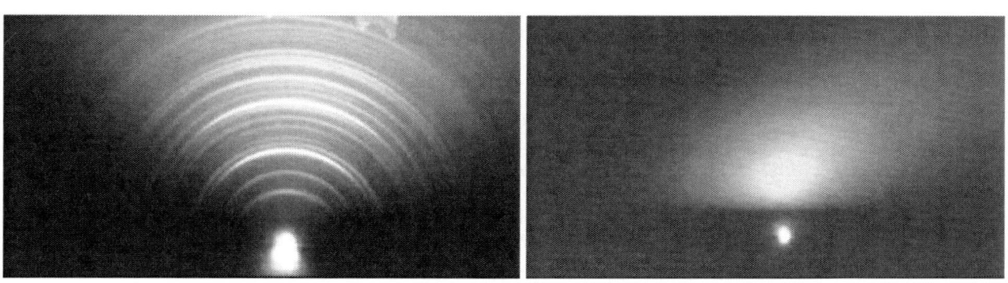

a) b)

Figure 3. Diffractograms of films obtained by RHEED from solutions based on systems: a) ZnO - SiO_2; b) SnO_2 - ZnO - SiO_2.

Figure 4 (a, b, c) shows that fractal dimension of structures decreases by increasing relative content of zinc oxide compared with relative amounts of tin dioxide. It may be noted that increasing of the maturation time is not change of this dependence.

Table 1. Calculation of fractal dimension of the surface in different ways

The ratio of SnO2-ZnO on the mole fractions	method name	150 hours	360 hours	1500 hours
2:1	decomposition	2,79	2,77	2,98
	counting cubes	2,51	2,38	2,60
	triangulation	2,60	2,39	2,69
	power spectrum	2,59	2,95	2,99
1:1	decomposition	2,82	2,72	2,70
	counting cubes	2,37	2,39	2,50
	triangulation	2,46	2,50	2,60
	power spectrum	2,54	2,48	2,16
1:2	decomposition	2,40	2,42	2,79
	counting cubes	2,18	2,17	2,60
	triangulation	2,15	2,16	2,69
	power spectrum	2,06	2,60	2,53

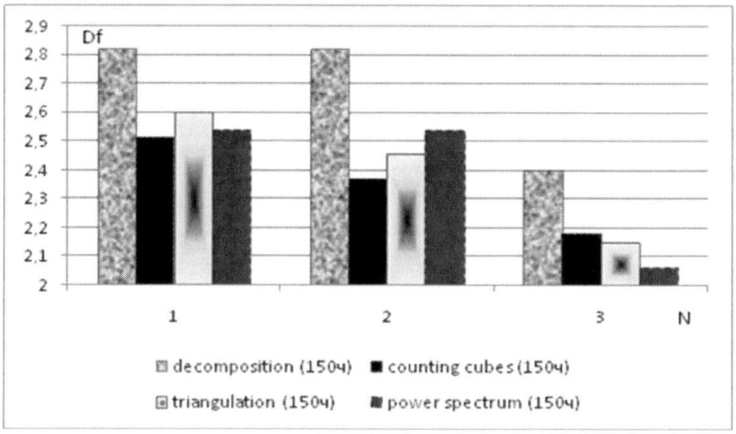

a)

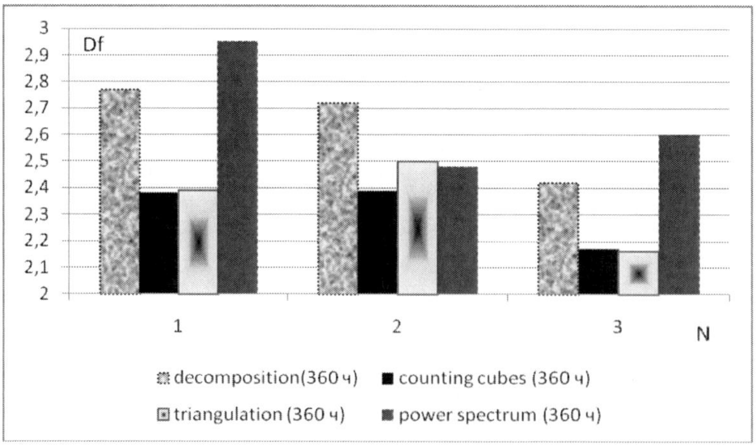

b)

Figure 4. (Continued).

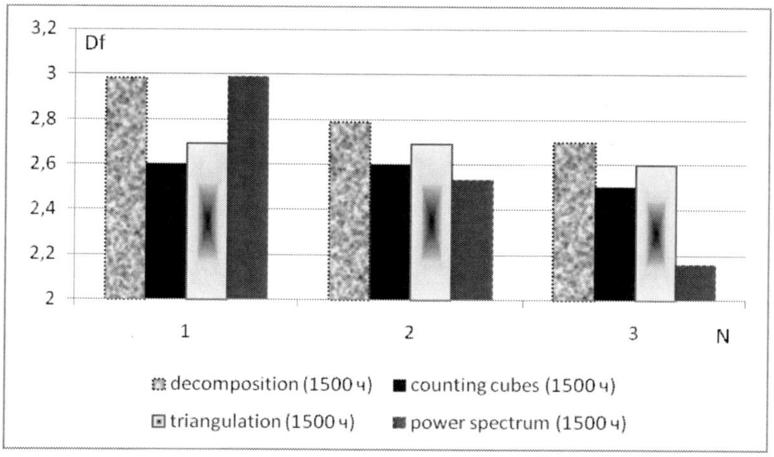

c)

Figure 4. The dependence of the fractal dimension (Df of SnO2 and a relative content of ZnO in the composition of the film 2:1, 1:1 and 1:2 for 1, 2 and 3 samples, respectively, the calculated different methods. At different times of ripening sol a) 150 hours, b) 360 hours, c) 1,500 hours.

It is known, that tin dioxide and silicon dioxide may to form long polymer chains and three-dimensional net structures, like [–Me-O-] , [–Me′-O-Me″-O-] and more intricates. Zinc oxide has two covalent bonds and is not form long polymer structures. Thus, we assume, that Zn blocks the development of three-dimensional polymer like –O– Me–O– from four valent atoms Sn and Si. Consequently, inclusion of ZnO in the growing polymer based on tin dioxide prevents the formation of the spatial chemical bonds in a certain local area and contributes to obtaining of a rarefied porous structure.

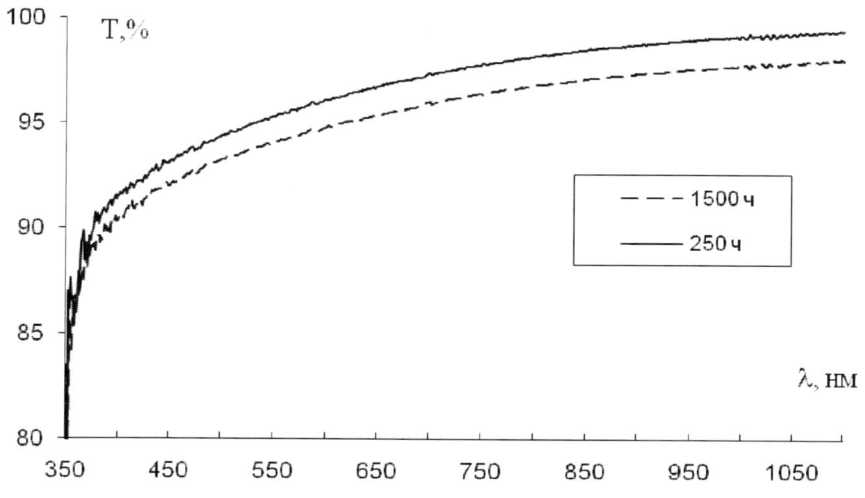

Figure 5. Dependence of transparent coefficient of samples on the radiation wavelength for different maturation time of the sol solution.

Figure 5 shows the dependence of the transmission coefficient T of thin films prepared on quartz glass by sol-gel from sols with different maturation time (250 h and 1500 h) from the wavelength λ. Analysis of the transmittance curves shown that transmittance coefficient of films decreased by an average of 2% during this period. The value T of the transmittance exceeds 89% for the shortwave light region of the spectrum, and 95% for the long wavelength. Thus, we can speak of a high optical bandwidth of the investigated films and their possible use in functional devices, requiring the use of multilayer optical structures.

Conclusion

As result of the investigation, influence of addition of ZnO on hierarchy of porous material based on SnO2 – SiO2 was detected. It has been shown that the interaction of zinc leads to amorphisation of the product of the sol-gel synthesis, based on tin dioxide and silica dioxide. It has been found that pore sizes can be controlled by varying the relative proportion of zinc oxide in a solution. These technological degree of freedom expends of the ability of the functional using of metal-porous materials, allowing to control the pore size by including of various modifiers in sol solution. Also quantified influence of the relative proportion of zinc oxide in a solution on the development of surface layers was obtained by calculating the fractal dimension. Obtained layers show high optical transparency.

References

[1] Chengxiang Wang, Longwei Yin*, Luyuan Zhang, Dong Xiang, Rui Gao/ Metal oxide gas sensors: sensitivity and influencing factors // *Sensors*, 2010, 10, p. 2088-2106.

[2] Moshnikov V. A., Gracheva I. E., Kusnezov V. V., Maximov A.I., Karpova S.S., Ponomareva A.A./ Hierarchical nanostructured semiconductor porous materials for gas sensors//*Journal of non-crystalline solids*, 2010, T. 356, №37-40, p. 2020-2025.

[3] I. E. Gracheva, V. A. Moshnikov, S. S. Karpova, E. V. Maraeva // Net-like structured materials for gas sensors / *Journal of Physics: Conference Series*. 2011. V. 291, № 1. P. 012017.

[4] Granqvist C.G., Hultaker A. / Transparent and conducting ITO films: new developments and applications// *Thing Solid Films*, 2002, V.411 , P.1-5.

[5] E. V. Abrashova., M.V. Baranovskiy // Preparation and Analysis of spectral characteristics of nanocomposites based on wide band gap semiconductors metal oxides of ZnO-SnO_2-SiO_2/ *Izvestiya of the St. Petersburg Electrotechnical University*, 2013. V. 5. p. 16-21. (in Russian).

[6] Pshchelko, N.S., Sensitivity Stability of Electret Transducers / N.S. Pshchelko, V.N. Tairov, F.R. Ozols // *Proc. ISE* 7. -1991. - C. 935-940.

[7] Pshchelko N.S., Polarization of the surface layers of ionic dielectrics on the border elektroadhesion contact with the conductor / *Nanostructured metals and materials* (ferrous metals). 2005. N 9. P.44-50.

[8] A. S. Bozhinova, N.V. Канева, I.E., Kononova et al. Photocatalite study and sensor properties of nanocomposite ZnO/SiO_2 layers. *Semiconductors*. V. 47, N 12. P. 1662-1666.

[9] Yanchinskaya E., Levine K.L., Tcaregorodceva M. and others // Preparing Films of Zinc Oxide With Varying Porosity by Sol-Gel Method // *Smart Nanocomposites* //V. 1, I 1 , p-55-70.

[10] Moshnikov V.A., Gracheva I.E., Pshchelko N.S., Anchkov M.G., Levine K.L. // Investigating Properties Of Gas-Sensitive Nanocomposites Obtained Via Hierarchical Self-Assembly // *Smart Nanocomposites* //V. 2, I 2, p-165-179.

[11] E. V. Abrashova, I.E., Gracheva, V.A. Moshnikov // Functional nanomaterials raised on metal oxides with hiererhical structure / *Journal of Physics: Conference Series*. 2013. T. 461. C. 012019.

[12] Levine K.L., Fen V.,. Nikonorova N.N and others // Studying Corrosion Protection with Nanostructured Polyimide Coatings by Electrochemical Impedance Spectroscopy// *Smart Nanocomposites* //V. 3, I 1 , p-49-58

[13] Brinker C. J., Scherer G. W. / *Sol-Gel Science*. The Physics and Chemistry of Sol-Gel Processing., San Diego: Academic Press, 1990. 908 p.

[14] Ed. Sumio Sakka, / *Handbook of sol-gel science and technology: processing, characterization, and applications*/. New York, 2004.–V.1-3.

[15] Moshnikov V.A,. Tairov Yu.M, Hamova T.V., Shilova O.A.// *Sol-gel Nanotechnology of Micro- and Nanocomposites* /, Saint-Petersburg: Lan', 2013, 304 p.

[16] R. Corriu, T.A. Nguyen // *Molecular Chemistry of Sol-Gel Derived Nanomaterials*. John Wiley & Sons, 2009.

[17] Inorganic polymers/*Encyclopedia of polymer science and technology*. v. 7.–N. Y.-L.-Sydney, 1967.–P.664–691,

In: Nanoscale-Arranged Systems for Nanotechnology
Editors: Kirill L. Levine
ISBN: 978-1-63482-353-1

Chapter 6

THE MECHANISM OF LITHIUM INTERCALATION INTO CARBON GRAPHITE POTLINING OF ALUMINIUM REDUCTION CELL

V. Yu. Bazhin*, A. G. Syrkov, R. Yu. Feshchenko and A. V. Saitov
National Mineral Recourses University, Saint-Petersburg, Russia

ABSTRACT

The problem of service life increasing of coal graphite materials in conditions of high temperature alumina cryolite melts is discussed in this article. The active influence of lithium in change of structural layers of graphite at the expense of their accretion as a result of intercalation has been established. Properties improvement of nanostructured carbon graphite material has been confirmed by series of the experimental tests which were carried out with various conditions.

Keywords: coal graphite nanomaterials materials, lithium intercalation, electrolyte, potlining

1. INTRODUCTION

Increasing of potlining durability is one of the important problems for extension in-service life of the aluminum electrolysis cell. During the operation of aluminium electrolysis cell alumina-cryolite melt is penetrating into some opened pinholes and cracks of the carbon-graphite material. So the interaction between some components begins, this process is accompanied with destruction of cathode lining into its bottom. The main reason for deterioration of potlining properties is sodium intercalation into layers of carbonaceous material which is adjacent to the melt that leads to swelling, loosening and forming gaps and

*Corresponding author. bazhin-alfoil@mail.ru.

micro cracks, thus the conditions for subsequent adsorption electrolyte and formation of carbide aluminum in blankets are improved.

It is known [1] that some expansion and deformation of many carbon materials is observed under the influence of sodium as the radius of sodium atoms is more than the distance between structural layers of the block from graphite material. As a result of sorption and interaction aluminium with carbon physical and chemical properties of a working layer of blocks are significantly changed and its link with the main structural basis is weakened therefore there is a destruction of the carbon layer, which is increasing in connection with a high temperature gradient of potlining as well as some temperature fluctuations in and shock influences [2].

On the other hand graphite can actively react with alkaline metals and pass into forms of connections (intercalated compounds) "alkaline metal-graphite" [3]. Sodium doesn't form any low-order connections NaC_x and passes to readily soluble forms of intercalated compounds with high atomic number. Similar interactions with carbon potlining are inherent also other alkaline metals and in particular lithium and potassium. Despite sufficient studying and a wide experience of practical operation of aluminum electrolysis cell with lithium electrolytes and some existing patents [4] the questions of interaction of lithium with carbon graphite material have not been covered enough in the literature [5].Some questions which are connected with formation of compounds of lithium and carbon have been studied but the mechanism of lithium penetration and the character its interaction in layers of carbon and graphite block has not been explained in the paper [2].

2. Experimental

It is known that so-called donor layered graphite compounds with alkaline metals are formed under processing of graphite by vapor of Li, Na, K, Rb, Cs or under fusion of graphite and alkaline elements [6]. Varying synthesis techniques (pressing, vacuum processing, annealing, use of monocrystals of graphite) and conditions of reactions so it is possible to receive graphite compounds of all stages with all alkaline metals [7].The investigations related to the behavior of lithium represent some scientific and practical interest because of the small sizes of lithium atoms which allow easy penetration with carbon layers. Therefore unlike other alkaline metals atoms of lithium are able to intercalate into the layers and block pin holes without any distortions of the carbon crystal structure. The reactions of lithium intercalation proceed with high speed in interlaminar space as carbon and graphite materials tend to form phases of lithium intercalation. An electric pit-type heating furnace has been used for conducting of experiments. The samples which had been cut out from industrial carbon graphite cathode blocks were placed in pendency in a tight steel glass. The temperature was controlled with platinum-rubidium contact thermocouples pressed in the pressure cap of the glass. The predetermined pressure of lithium vapors was set up with gate regulation. The scheme of experimental installation is shown in Figure 1.

Initial characteristics of the samples are presented in Table 1. Lithium carbonate was used as an initial raw material. After heating Li_2CO_3 up to the temperature 750 ° C in the electric pit-type heating furnace there is a formation of lithium oxide which is partially sublimated at temperature above 1000 °C according to the following reaction.

$$Li_2CO_3 \rightarrow Li_2O + CO_2 \quad (1)$$

The process of deoxidation of lithium oxide (by using crystal silicon in the form of small waste) was applied to impregnate carbon graphite material for receiving lithium vapors. A layer of silicon dust was covered evenly with a layer of lithium carbonate.

Steady oxide of silicon is formed by interaction of Si with Li_2O sublimates.

$$2Li_2O + Si = 4Li + SiO_2 \quad (2)$$

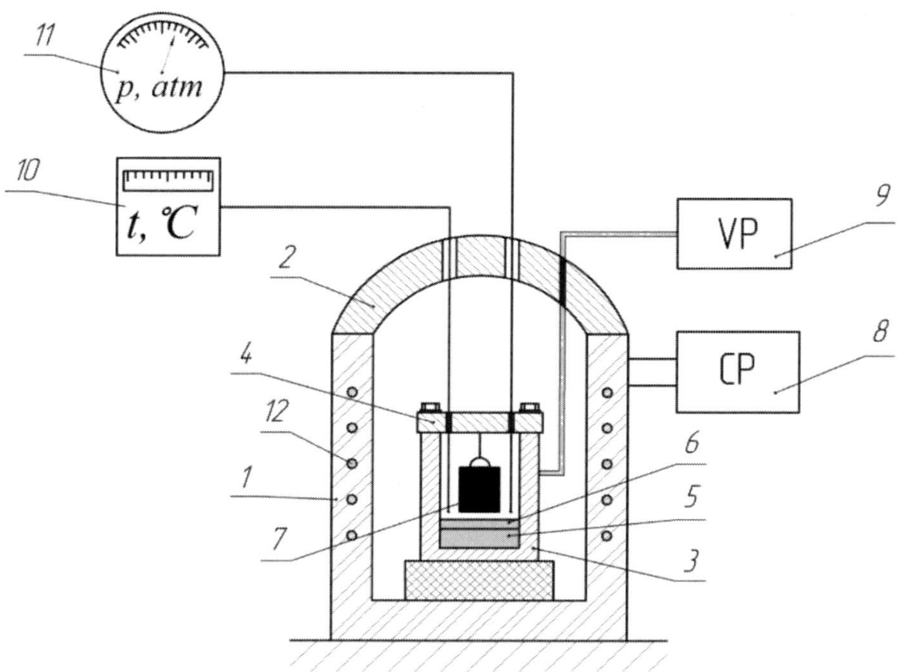

Figure 1. Scheme of experimental installation: 1 – electric pit-type heating furnace; 2 – pressure cap; 3 – steel pad; 4 – steel beaker; 5 – lithium carbonate; 6 – silicon; 7 – graphitized carbon material; 8 – control panel; 9 – vacuum pump; 10 – platinum rhodium thermocouple; 11 – combined pressure gage; 12 – chrome wire.

After the fulfilled experiments have been carried out, five samples were tested and all changes of their properties were analyzed. Characteristics of carbon graphite materials before impregnation by lithium and after mathematical calculation are presented in Table 1.

Table 1. Average properties of carbon materials

Materials	Electric resistance, mOm (20 °C)	Thermal conductivity, W/K·m	Ash content, (%)	Porosity (%)
Initial carbon material	35	8	2,5	14
Lithium covered samples	17	35	0,35	21

The character of the measured adsorption surface and the surface with high apparent density are the standards for the adsorption researches on an electron microscope. The electron microscope TESKAN with configuration VEGA (x10000) was used for the most complete understanding of the mechanism of lithium accumulation in the coal material. The structure of the materials was studied by X-ray diffraction (XRD) analysis using a XRD-7000S (Shimadzu) diffractometer with Cu Kα (λ=1.5405 Å) X-ray source.

3. Results and Discussion

The executed research showed that at first lithium penetrates into more porous parts of material, therefore the diffusion mechanism of its intercalation into layers of the carbon graphite sample is more likely. Depth of penetration of lithium can be determined when cutting a cylindrical sample after experiment and the subsequent electronic microscopic research (Figure 2).

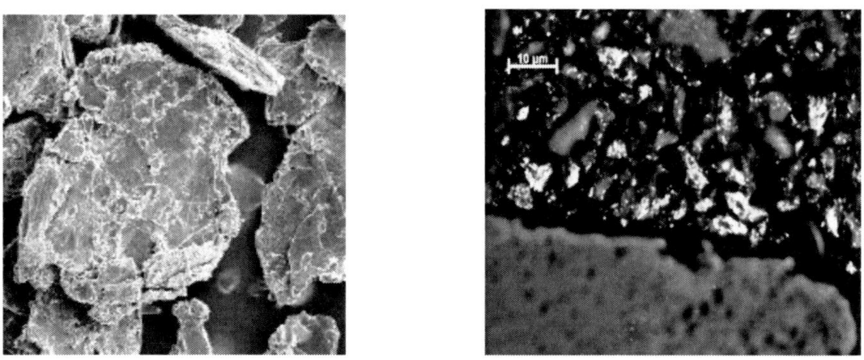

a *b*

Figure 2. Analysis of sections executed by means of the scanning electronic microscope in the internal area of the coal sample: *a* – at the beginning of lithium metallization; b – formation of intercalation Li_xC_y (x1000).

On the scanned sections of the electron image (Figure 2,*a*) the contours of a rectangular layered particle of anthracite can be seen. The executed in XRD patterns showed that only lithium has rather deep intercalation (5-15 mm) and thus its atoms are not transported on the borders of grains. Most likely this is elementary lithium because the emission of X-rays was not increased in comparison with the initial state of the material. The area of a porous matrix of binding carbon which is situated between dense particles can be also visible in the picture. Defined in X-rays lithium emission is observed in the all parts of material but its significant increase has been revealed in porous structure. As a result the research has shown that, despite the fact that the penetration of lithium extends across the crystal lattice, it is limited to migration into more porous matrix of the binder component.

The lithium movement within carbon material can be considered as a diffusion driven process which derived from the source (volume) with a constant concentration. The solution of the equation using Fick's law, in this case gives [8]:

$$\frac{C_i - C_x}{C_i - C_0} = \frac{2}{\sqrt{p}} \cdot \int_0^k \exp\left(-y^2\right)dy \tag{3}$$

where

$$k = \frac{x}{2 \cdot \sqrt{D \cdot t}} \tag{4}$$

C_i – concentration particles at diffusion limit of the section x=0 which during penetration is a constant;

C_x – concentration with a depth of diffusion x and t time;

C_o – concentration (a constant at any value x) at the beginning of diffusive process.

The research has determined that there is a formation of intercalated compounds Li_xC_y, at the temperature 1350-1400 °C at the interface of lithium penetration with the micro pin holes. Evidently, lithium adsorption, which occurs between layers and (to its extent) is related to defects in the crystal structure of graphite which serves as the primary source for lithium accumulation mechanism. The main argument in favor of the adsorption and interaction of lithium intercalated compounds within the graphite layers is able to serve a shrinkage which is an evident of disintegration of some layers and volume reduction of the samples by 5-7%. Stages of the mechanism of interaction of lithium with carbon and graphite materials:

1 stage – lithium adsorption on a surface in a time and cracks and primary introduction of lithium in the graphite layers with separate atoms at the expense of excessive pressure;
2 stage – uniform distribution of atoms in volume of the samples;
3 stage – process of intercalation of lithium with carbon;
4 stage – accretion of carbon layers.

Changes in the structure of graphite after lithium intercalation is shown in Figure 3.

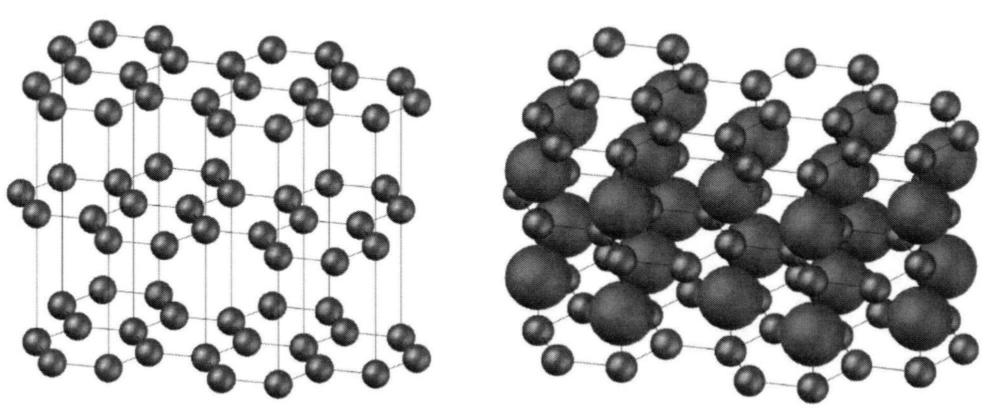

a *b*

Figure 3. The mechanism of lithium intercalation with carbon-graphite materials: *a* – primary consist; b – after processing by lithium.

It should be noted that the migration of most lithium intercalated compounds (LiC_6, Li_2C_8, Li_6C_{32}) takes place between graphite layers in its crystallites and extends immediately directly for all layers, i.e., perpendicular to the graphite layers, which leads to their coalescence. This fact confirms Ubbelohde theory [9] of halogen intercalation into graphite layers. Some sections of deformation of the crystallite boundaries occur during partial penetration into the layer. Small distortions of the crystallite boundaries will generate more significant changes in the structure of graphite which primarily is related to its hardening. If each of these models represent a separate modular particle that the scheme of disintegration shown in Figure 3 is the most probable provided that the distortion is increasing and sufficient to change the crystallite boundaries for appropriate carbon graphite material durability.

Thickness of the filled layer is identical to all stages but depends from nature of metal and for lithium changes within 0.45-0.55 nanometers. All these interreactings are constructed equally due to ionic interaction of a cation of Li^+ and macro carbon anion nC^- because of electron transfer from an external orbit of C in graphite grid. In favor of such model testifies the conductivity increase (Table 1) upon transition of graphite to an intercalate layer and even emergence of superconducting properties [6].

Besides the intercalation filled distance to equally doubled sum of radiuses of Li^+ and of Van-der-Vaals thickness of carbon layer. The effect of accretion of graphite layers is shown in Figure 4.

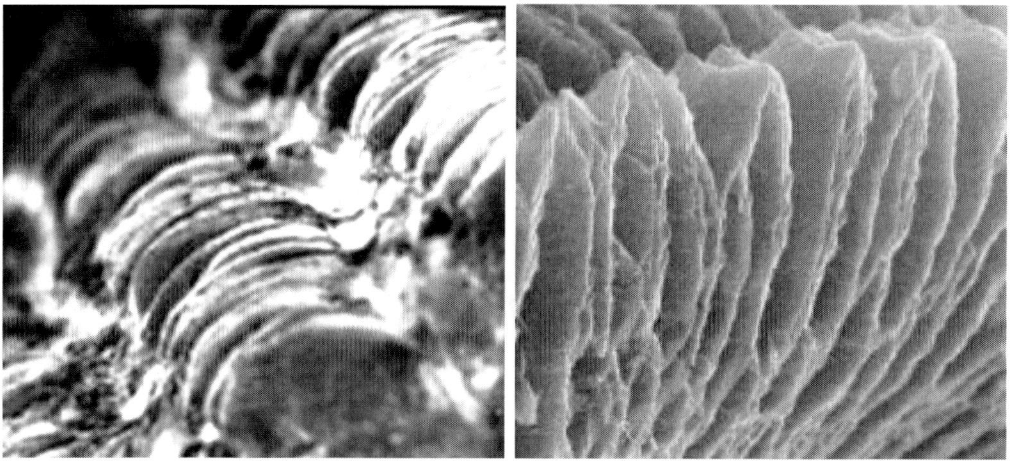

Figure 4. The effect of accretion of graphite layers.

Lithium vapors diffuse through the carbon potlining where they are inhibited by some adsorbing layers of graphite and then they are condensed in the cold parts of installation. Lithium atoms react with carbon to form intercalated compounds and to accrete layers when there is a migration in a high temperature range 1300-1500°C. The efficiency of the process of lithium intercalation into the carbon graphite materials depends on the condition of block surface and its structure, which determines the kinetics of the intercalation process and the amount of intercalated lithium. Initial characteristics of carbon graphite material are sharply changing at a high reaction rate in this case there is a significant importance in the composition and structure of the filler particles acting as a matrix to dissolve lithium.

The atomic-emission research of samples obtained after treatment with lithium at different temperature and time intervals allows defining the mechanism of intercalation of lithium atoms into the carbon graphite material. As a result, it has been proven that there is a lithium intercalation and pin holes metallization at the first stage of heat treatment of samples with vaporized lithium. The samples were cut out from industrial carbon blocks. Compounds like Li_xC_y were formed and collected in the surface coatings with the passage of time at elevation of temperature. The compounds are accumulated in the block depth between graphite layers. The steady phase LiC_6 is formed during further intercalation and lithium penetration. It is defined that the energy of activation of these phases formation makes respectively 12,4-12,6 kJ/mol. Initial concentration of lithium defects in the temperature range 1300-1500 ° C makes 0,005-0,014 mol/cm^3, diffusion coefficient (9,76 ±0,5) sq·m/sec.

Conclusion

The fulfilled research showed that accumulation of lithium atoms in the layers of carbon graphite has steady character. Under the same conditions of the experiment most part of lithium is adsorbed within the layers of anthracite 5-7 mm and 8-10 mm graphite. Thus the structure and some properties of the surface layer of carbon graphite materials are changed. Lithium concentration reaches a certain level in the surface layer, and the main part of the lithium is able to keep the original order and structure of the carbon graphite materials without any distortions of horizontal structure of layers.

Some changes in the structure of individual crystallites between the layers of graphite could be the next step or the sum of the structural changes which lead to a fundamental change of properties of coal-graphite materials such as open porosity, conductivity and heat conductivity. In turn the metallization of carbon structure by atoms of lithium provides a unique possibility to create carbon graphite materials of a new generation with improved characteristics and opportunity to accumulate energy in cathode lining for example of reduction aluminum cell.(in the surface layers)

References

[1] M.B. Rapoport, V.N. Samoilenko, «Deformation of cathode blocks in aluminum baths during process of electrolysis», *Tsvet Met* 30 (1957). P. 44-51.

[2] I. Pflugmacher and H.P. Boehm, 3rd Conf. Ind. *Carbons Graphite* (1970).

[3] Morten Sorlie, Harald A. Oye, *«Cathodes in aluminum electrolysis»*, 3rd edition, (2010), 217-218.

[4] Patent 102304726 (A) CN C25C3/12 01.04.2012. (USA) «Aluminum electrolytic prebaked anode and preparation method thereof». 01.04.2012.

[5] G.N. Makarov, G.D. Kharlampovich, «Chemical technology of solid fossil fuels». College textbook. – Moscow: *Chemical*, 1986. 496 pp. (in Russian)

[6] Zabel, H. Intercalated graphite / H. Zabel, Chow P.C. // *Comments Condensed Matter. Phys.* – 1986. – Vol. 12, № 5. – P.225-251.

[7] McDonnel, F.R.M. Some physical properties associated with "aromatic" electron. Part III. The pseudo-metallic properties of potassium-graphite and graphite-bromine /

F.R.M. McDonnel, R.C. Pink, A.R. Ubbelohde // *J. Chem. Soc.* – 1951. – N.1. – P.151-197.

[8] I.M. Gluschenko, «*Theoretical Foundations of Technology of solid fossil fuels*». – Kiev: High school, 1980. 255 pp. (in Russian)

[9] Ubbelohde A.R., Lewis F.A. *Graphite and its crystal compounds*. Oxford (1960). 249 pp.

In: Nanoscale-Arranged Systems for Nanotechnology
Editors: Kirill L. Levine
ISBN: 978-1-63482-353-1

Chapter 7

Synthesis of Cadmium Selenide Colloidal Nanoparticles Stabilized with Thioglycolic Acid

D. S. Mazing* and O. A. Aleksandrova
Saint Petersburg Electrotechnical University "LETI", Russia

Abstract

Colloidal nanoparticles of cadmium selenide with average size ranging from less than 1,5 nm to 2,2 nm were obtained in water medium using thioglycolic acid (TGA) acting as surfactant. Series of experiments with different component molar ratios and temperature regimes was performed in order to reveal their characteristics and size dependence on technological parameters. Characterization methods include absorption and photoluminescence spectroscopies, X-ray diffraction analysis. Synthesized nanocrystals might be used for biomedical detection.

Introduction

Semiconductor colloidal quantum dots (QDs) have recently attracted much attention because of their tunable optical properties and technological flexibility. Due to strong quantum confinement, cadmium selenide nanoparticles exhibit luminescence in almost entire visible range, depending on their size. One of the emerging fields for their application is biological and medical detecting [1] where QDs may rival conventional organic fluorophores. In comparison with organic fluorophores semiconductor nanocrystals are much more stable to photodegradation. Moreover absorption band wideness and variety of available emission colors make it possible to excite a set of QDs with various photoluminescence peak positions with the help of the same radiation source, thus allowing detecting signals from different targets simultaneously.

*E–mail: dmazing@yandex.ru.

Since hot – injection method was proposed [2], plenty of methods of colloidal quantum dots synthesis were developed aimed to reduce cost and hazards associated with usage of toxic and pyrophoric compounds. One of the natural choices of medium for QDs preparation might be polar solvents, for example water. Despite poorer crystallinity and therefore generally inferior optical properties, as compared to nanoparticles synthesized in non-polar media, these methods tend to be more environmentally benign and provide nanocrystals compatibility with biological tissues, eliminating necessity for ligand exchange operations. Suitable precursors in this case may include water-soluble metal and chalcogenide salts as sources of metal and chalcogenide ions, respectively, while thiols might be used for surface passivation.

MATERIALS AND METHODS

$NaBH_4$, $CdCl_2{\cdot}2.5H_2O$, elemental Se (powder), thioglycolic acid (TGA, 80%), KOH (6M), propanol-2 (anhydrous) were used without any further purification. Distilled water was used as solvent.

Absorption spectra were acquired on a PE-5400UV UV–vis spectrophotometer (LLC "Ekohim"). Photoluminescence measurements were carried out on equipment based on fast scanning spectrometer Ocean Optics USB-4000 using laser excitation on λ = 445 nm. XRD analysis was performed with diffractometer DRN "Farad" using CuKα radiation.

In order to determine average size of the particles polynomial expression [3] was used:

$$a=\left(1.6122\cdot10^{-9}\right)\lambda^{4}-\left(2.6575\cdot10^{-6}\right)\lambda^{3}+\left(1.6242\cdot10^{-3}\right)\lambda^{2}-0.4277\lambda+41.57, \qquad (1)$$

where a is average size of particles, λ is a wavelength of the first excitonic peak. This equation was obtained by the procedure of interpolating transmission electron microscopy data for CdSe nanoparticles from various sources and it is considered to give more reliable assessment than effective mass model approximation does [3].

EXPERIMENT AND RESULTS

Synthesis procedure included following steps. Sodium hydroselenide solution was prepared by dissolving $NaBH_4$ and selenium powder in distilled water in a glass vial. It resulted in vigorous reaction accompanied with intensive hydrogen gassing. The reaction might be described as follows:

$$4NaBH_4 + 2Se + 7H_2O \rightarrow 2NaHSe + Na_2B_4O_7 + 14H_2$$

To obtain cadmium precursor solution cadmium chloride hydrate was dissolved in distilled water and certain amount of TGA was added so that the mixture became turbid. By dropwise addition of KOH (6M) acidity of the solution was brought up to approximate value pH = 11 and it became clear. Such moderate alkaline environment facilitates formation of Cd^{2+}-TGA complexes which act as immediate sources of cadmium ions.

In one implementation the molar ratio of the components was $[Cd^{2+}]:[Se^{2-}]:[TGA]=1:0.5:2.4$. Freshly prepared NaHSe precursor solution was swiftly injected into cadmium containing solution at 25 °C which resulted in rapid nucleation indicated by color change. The first probe (1-1) was taken immediately after nucleation. Then swift heating was provided and aliquots were taken at different temperatures: 1-2 – 35°C, 1-3 – 50°C, 1-4 – 70°C. Corresponding absorption and photoluminescence spectra are depicted on Figure 1 and Figure 2 respectively.

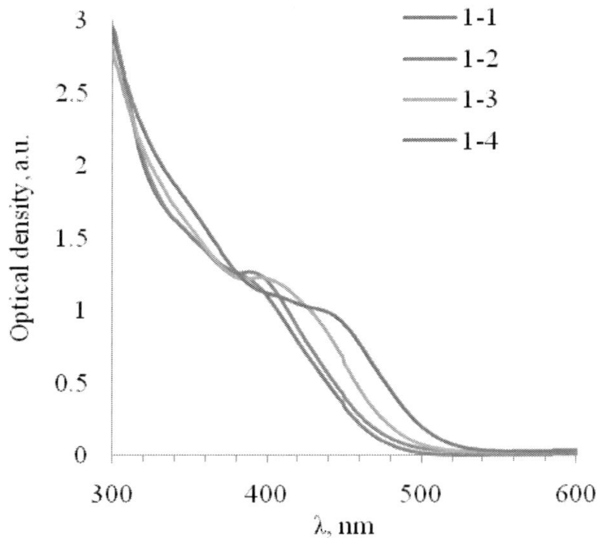

Figure 1. Absorption spectra of CdSe nanoparticles prepared with molar ratio $[Cd^{2+}]:[Se^{2-}]:[TGA]=1:0.5:2.4$.

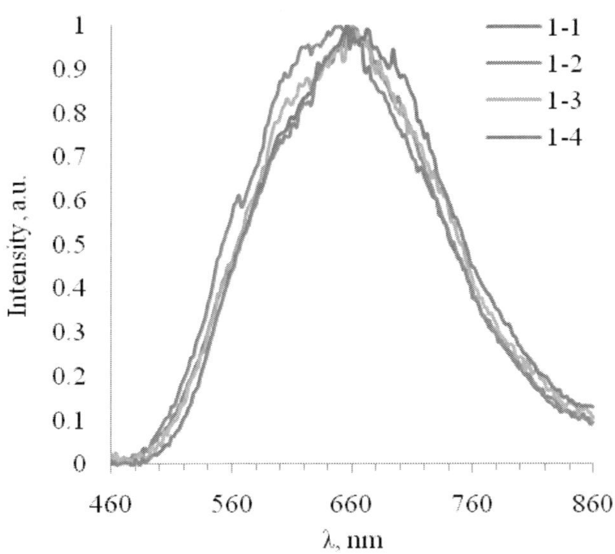

Figure 2. Photoluminescence spectra of CdSe nanoparticles prepared with molar ratio $[Cd^{2+}]:[Se^{2-}]:[TGA]=1:0.5:2.4$.

It is seen on the Figure 1 that absorption onset shifts to larger wavelength during the reaction which indicates continued nanoparticle growth. The second sample has the most distinct first excitonic maximum while consequent probes are characterized by broader first transition with tendency to amplitude decrease. Such absorption spectra behavior implies particle monodispersity deterioration during the synthesis. Photoluminescence spectrum doesn't change its position and features broad peak with large wavelength offset between first absorption transition and emission band maximum which is called Stokes shift. It means that trap states on the surface of nanocrystals account for radiative recombination which may occur because of an insufficient passivation. In another implementation molar ratio $[Cd^{2+}]$:$[Se^{2-}]$:[TGA]=1:0.25:1.2 was used. This time NaHSe precursor solution was intensively injected in cadmium containing solution at 45 °C and temperature was gradually raised throughout the reaction. Aliquots were taken consequently in following intervals: 2-1 – 0.5 min (50 °C), 2-2 – 6 min (51 °C), 2-3 – 22 min (58°C), 2-4 – 60 min (63 °C), 2-5 – 90 min (64 °C), 2-6 – 125 min (95 °C). Resulting spectra are shown in the Figures 3 and 4.

Compared to the first synthesis, this experiment has yielded smaller nanoparticles. Expression (1) gives approximate average size of QDs less than 1.6 nm (for the sample 2-6). As can be seen on absorption spectra of the first two samples, the peak at wavelength of 368 nm doesn't change its location for at least 6 minutes after injection. Therefore, considering size assessment, it might be assumed that thermodynamically stable magic-sized clusters were formed on nucleation stage. This assumption was supported by next similar experiment, when absorption feature on the very same wavelength was observed. Absorption curves retain the first excitonic transition maximum amplitude, prompting uniform growth and relatively stable particle monodispersity during the reaction process. Thus obtained nanoparticles exhibit broad band emission and large Stokes shift as well, with the peak maximum shifting in the range from 580 to 640 nm in accordance with absorption onset. It is also notable that temperature increase resulted in higher growth rate.

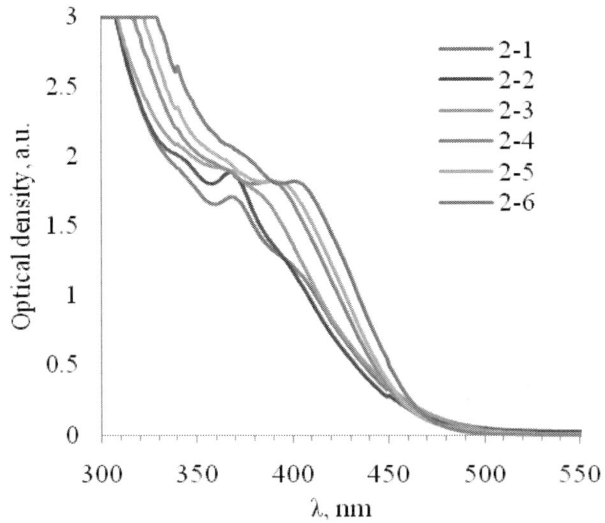

Figure 3. Absorption spectra of CdSe nanoparticles prepared with molar ratio $[Cd^{2+}]$: $[Se^{2-}]$:[TGA]=1:0.25:1.2 and injection temperature 45°C.

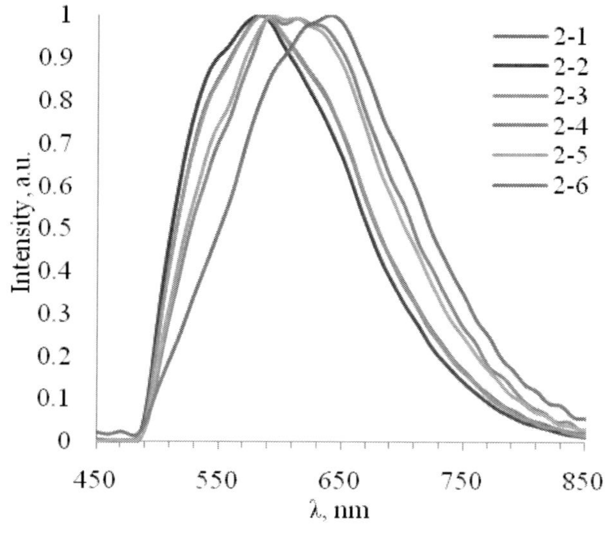

Figure 4. Photoluminescence spectra of CdSe nanoparticles prepared with molar ratio $[Cd^{2+}]:[Se^{2-}]:[TGA]=1:0.25:1.2$ and injection temperature 45°C.

Another synthesis was implemented at higher injection temperature, so that precursors were mixed at 97 °C and this value was maintained constant. Samples were consequently acquired in 0.5; 6; 25; 60 and 90 minutes after start of the reaction. Molar ratio was $[Cd^{2+}]:[Se^{2-}]:[TGA]=1:0.6:1.4$. Absorption and photoluminescence spectra are shown on Figure 5 and Figure 6.

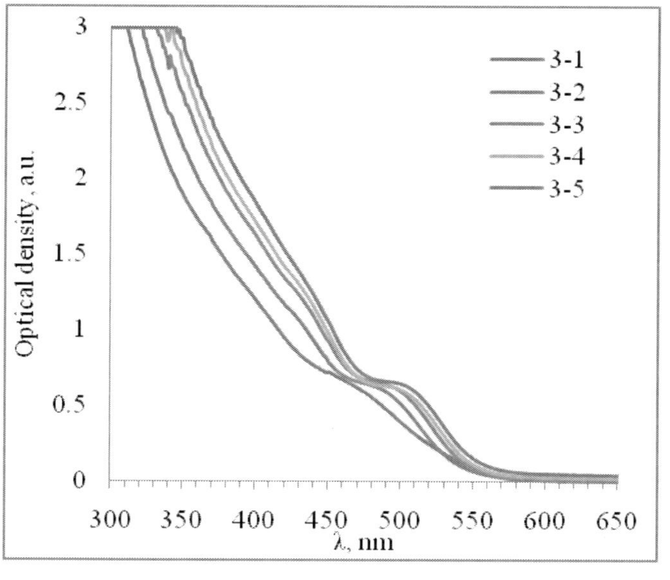

Figure 5. Absorption spectra of CdSe nanoparticles prepared at 97°C.

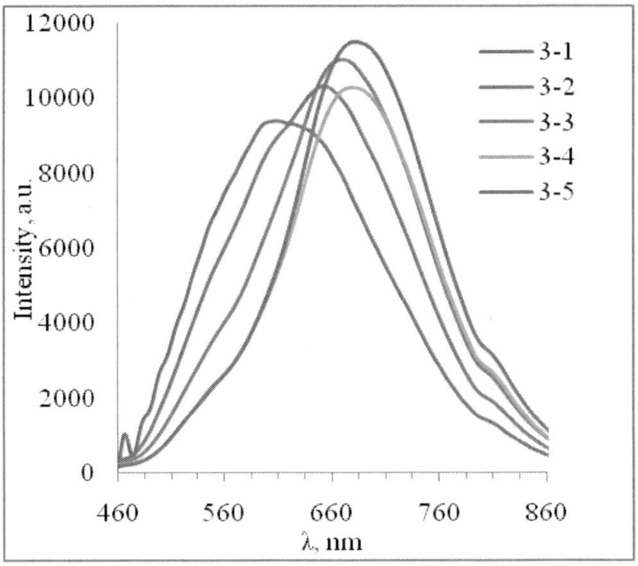

Figure 6. Photoluminescence spectra of CdSe nanoparticles prepared at 97 °C.

As can be seen from absorption spectra (Figure 5) comparatively larger nanocrystals were formed on nucleation stage and growth rate slowed down in approximately 25 minutes. In this case average size of the particles estimated from first excitonic peak wavelength values by means of expression (1) was less than 2.2 nm. Photoluminescence peak maximum shifts from 615 to 683 nm (Figure 6) and as in previous experiments it is broad and exhibits large Stokes shift. In order to define QDs crystal structure X – ray diffraction analysis was performed on the fourth sample of the second series (2–4) transformed into a powder through addition of propanol-2 with subsequent centrifugation.

Generally cadmium selenide may crystallize in two forms: cubic zinc blend and hexagonal wurzite. In our case the first peak on the diffractogram (Figure 7) likely corresponds to the $(10\bar{1}0)$ plane of wurzite crystal structure. Its noticeable asymmetry may be attributed to overlapping reflections from (0002) and $(10\bar{1}1)$ planes which generally are less intensive. In this case the widened reflection around $2\theta = 64$ degrees should be ascribed to $(11\bar{2}0)$ plane.

As a result it may be stated that examined sample likely has hexagonal wurzite structure which is unusual for CdSe nanoparticles prepared in water solutions at moderate temperatures [4–6]. Nevertheless taking into account very small average size of the particles the possibility of phase coexistence shouldn't be excluded. Not extracted nanocrystals stayed stable in solution for approximately 1 – 1.5 months after synthesis without any special measures for sample storage. It is presumed that thiol stabilized nanoparticles are rather sensible to photooxidation. Proposed mechanism consists in formation of disulphides on the surface and bond rupture between ligands and crystal surface. In this case particles facets itself are exposed to etching because of the reaction between oxygen and selenium with selenium oxide formation. This leads to decrease of nanocrystal size, and, as consequence, luminescence shifts toward shorter wavelengths. At the same time organic layer degradation brings about sample precipitation [6]. Our samples exhibited very similar behavior changing luminescence color from yellow to green with concurrent precipitation.

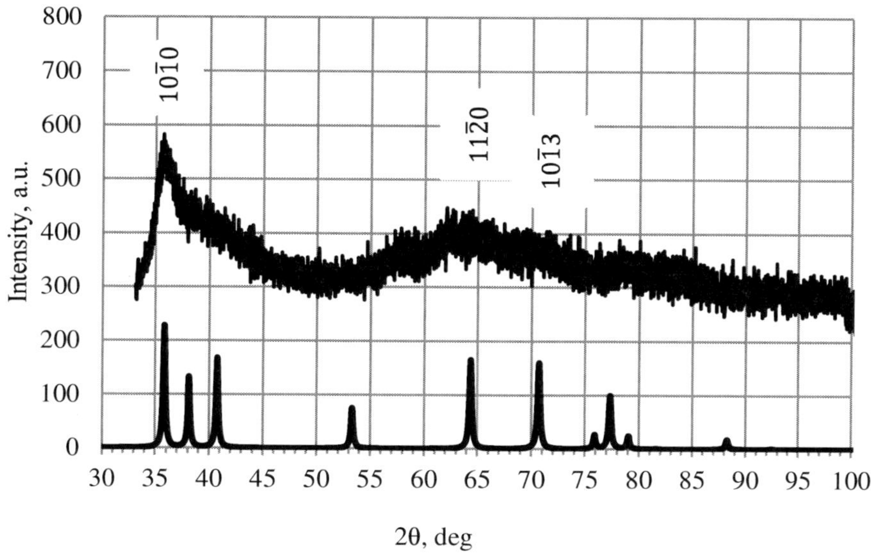

Figure 7. X-ray diffraction analysis of nanocrystals CdSe 2-4 and theoretic reflexes of wurzite phase.

CONCLUSION

In this work synthesis of CdSe nanocrystals with average size less than 2.2 nm was implemented through wet chemistry technique using thioglycolic acid acting as surfactant. Nanoparticles exhibit photoluminescence with maximum peak in the range between 580 – 680 nm depending on technological parameters. Photoluminescence is characterized by broad peaks and large Stokes shift indicating that trap states play major role in emission process. Precursor injection at higher temperature results in formation of relatively larger nanocrystals on nucleation stage, while raising temperature during the process of the reaction leads to more intensive growth. X-ray diffraction analysis indicates that prepared QDs have hexagonal wurzite structure.

REFERENCES

[1] Walling M. A., Novak J. A., Shepard J. R. E. *Int. J. Mol. Sci.* 2009 10 441–491.

[2] Murray C. B., Norris D. J., Bawendi M. G. *J. Am. Chem. Soc.* 1993 115 8706–15.

[3] Yu W. W., Qu L. H., Guo W. Z., Peng X. G. *Chem. Mater.* 2003 15 2854–60.

[4] Sun Q., Fu S., Dong T., Liu S., Huang C. *Molecules* 2012 17 8430–38.

[5] Sharma P. K., Dutta R. K., Liu C. H., Pandey R., Pandey A. C. *Mat. Lett.* 2010 64 1183–86.

[6] Rong X. L., Zhao Q., Tao G. H. *Chin. Chem. Lett.* 2012 23 961–964.

[7] Kloepfer J. A., Bradforth S. E., Nadeau J. L. *J. Phys. Chem. B.* 2005 109 9996–10003.

In: Nanoscale-Arranged Systems for Nanotechnology
Editors: Kirill L. Levine
ISBN: 978-1-63482-353-1

Chapter 8

Investigation of the Optical Properties of Nanoporous Membranes Based on Alumina

Ekaterina N. Muratova* and Lev B. Matyushkin
Saint Petersburg Electrotechnical University "LETI", Russia

Abstract

Membranes of nanoporous aluminum oxide (alumina) have been obtained using electrochemical etching technique by varying of technological regimes. The surface morphology and cross-sections of experimental samples are studied using scanning electron microscopy (SEM). The optical transmission measurements in wide wavelength range of 190–12000 nm were performed. It is hypothesized the average pore diameter and the pore diameter dispersion can be determined by UV-visible spectra while modification of alumina, membrane thickness and the type of electrolyte anions embedded into the porous structure can be determined by measuring near and mid-infrared spectra.

Keywords: nanoporous alumina membranes, transmission spectra, structure parameters

Introduction

Porous alumina due to the unique structure is increasingly studied to be able to expand its areas of application. The electrochemically produced porous alumina (por-Al_2O_3) membranes have been widely used as templates for the fabrication of ordered nanostructures [1-3], owing to regular nanoscale structures and relatively easy and low-cost processing. Of considerable interest is the por-Al_2O_3 for the problems of biomedicine [4], various types of sensors [5] and MEMS-devices [6, 7]. A special attention is paid to the study of the optical properties of both the porous structures [8] and various composites based on them [9]. Due to their structure, these membranes have potential in various optical device applications, such as optical filters

*Corresponding author: Saint Petersburg Electrotechnical University "LETI", ul. Prof. Popova, 5, 197376 Russia. Tel.: + 7 921 388 69 65. E-mail: SokolovaEkNik@yandex.ru.

and photonic crystals [10]. It is known that aluminum anodized in acidic electrolytes may occur at different rates, efficiency and stability, in general, depend on the nature and concentration of the electrolyte, temperature and anodic current density [11].

It is important to be able to give a quick characteristic of the porous structure parameters to evaluate the average pore diameter, size distribution and order. Thus, the aim of this work is to determine the effect of porous alumina membrane structural parameters on the transmission spectrum.

EXPERIMENT AND RESULTS

In order to obtain a porous anodic alumina membrane (Figure 1) electrochemical anodization of aluminum foil was fixed in special Teflon frame with an opening area of 7 mm^2 and then it was conducted in potentiostatic mode in electrolytes based on aqueous solutions of phosphoric (H_3PO_4) or sulfuric (H_2SO_4) acids. In order to slow the etching process and improve the quality of the membrane 10–15% glycerol was added to the electrolyte.

Figure 1. Photo of alumina membrane.

Usually the investigation of the structure parameters (diameter of pores, interpore distance, porous layer's thickness) of samples is carried out with help of scanning electron microscopy (SEM). Anodic alumina membranes contain many columnar pores with diameters of about 20-200 nm (Figure 2). Geometric parameters of membranes obtained in different electrolytes are presented in Table 1.

The optical transmission measurements were performed on PE-5400UV spectrophotometer (LLC "Ekohim") in the wavelength range of 190–1000 nm, SF-256BIK (LLC "LOMO") in the wavelength range of 1000–2700 nm and FTIR spectrophotometer in the wavelength range of 2600–12000 nm.

The study of UV-visible transmission spectra revealed that for any membrane the transmission spectrum has the same shape (Figure 3): transmission of the membrane starts to rise only at a certain critical wavelength λ_c. A comparison of transmission spectra for the different membranes (Figure 4) shows that λ_c depends on the type of electrolyte in the presence of which it was produced. The band gap of alumina is too large (according to various literature 7,0 – 9,5 eV) to observe the size effects in the available optical range.

This phenomenon can be related to a) the pore size or b) the type of anion incorporated into the anodized alumina lattice (SO_4^{2-}, PO_4^{3-}). However, since it is anion type determines the pore size, we can assume that both of these explanations are interrelated. It was found the following pattern: the smaller the pore diameter, the smaller the critical wavelength λ_c, also the more ordered structure, the larger slope of the curve. It is important to note there is always a range of the pore diameter, which could be described by a Gaussian function (Figure 5).

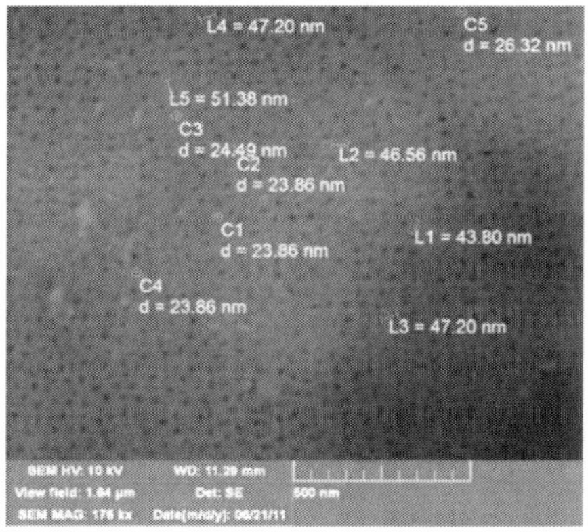

a

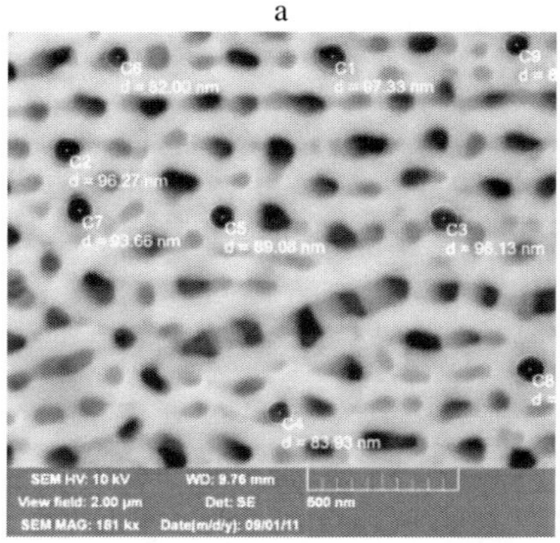

b

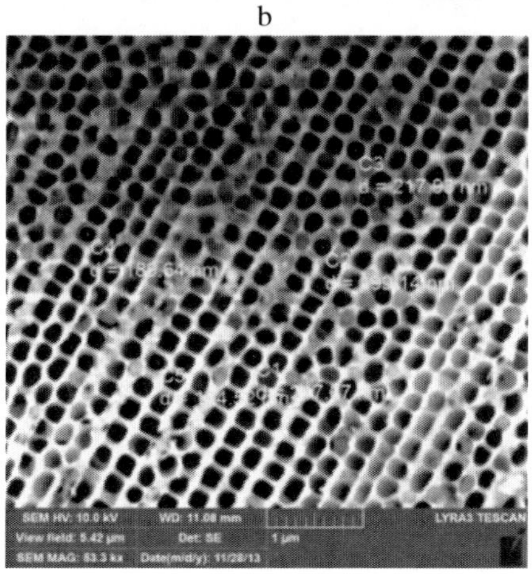

c

Figure 2. SEM images of the membranes por-Al_2O_3 obtained in an electrolyte based on solutions of H_2SO_4 (a) and H_3PO_4 (c, b).

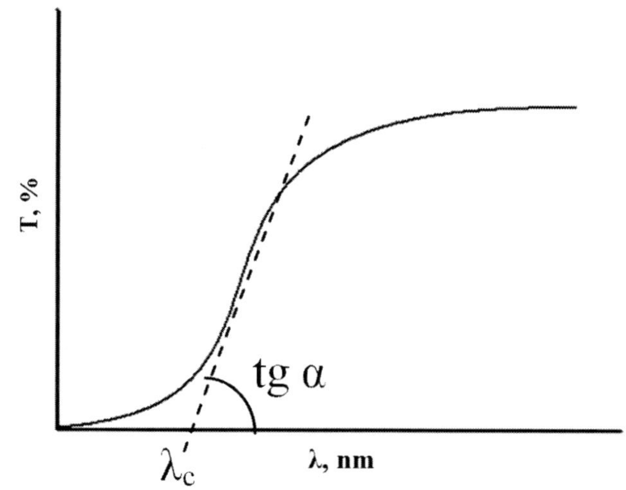

Figure 3. General view of porous alumina UV-Vis transmission spectra.

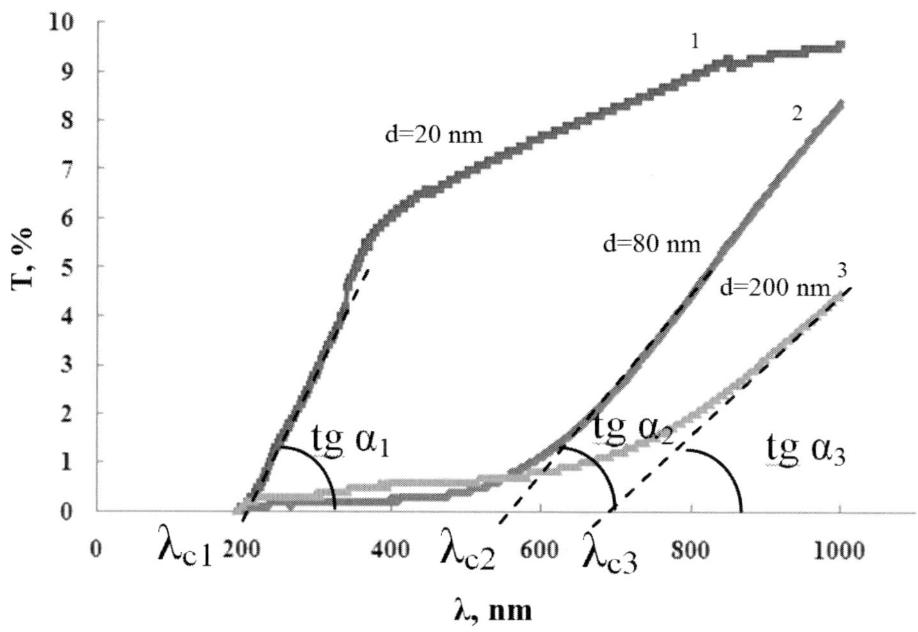

Figure 4. Transmission spectra of porous alumina membranes produced in presence of 1 – SO_4^{2-} and 2, 3 – PO_4^{3-} anions.

Analysis of experimental data shows some correlation between λ_c and type of porous structure due to the type of electrolyte. Summary of the experimental data presented in Table 1. Thus, the presented generalization of experimental data allows determining the average size and the dispersion of pore diameter by UV-visible transmission spectrum measuring.

Figure 5 shows mid-IR transmittance spectra for membranes obtained in electrolytes based on aqueous solutions of sulfuric (Figure 6 (a)) and phosphoric (Figure 6 (b)) acids with 10–15 vol. % of glycerin.

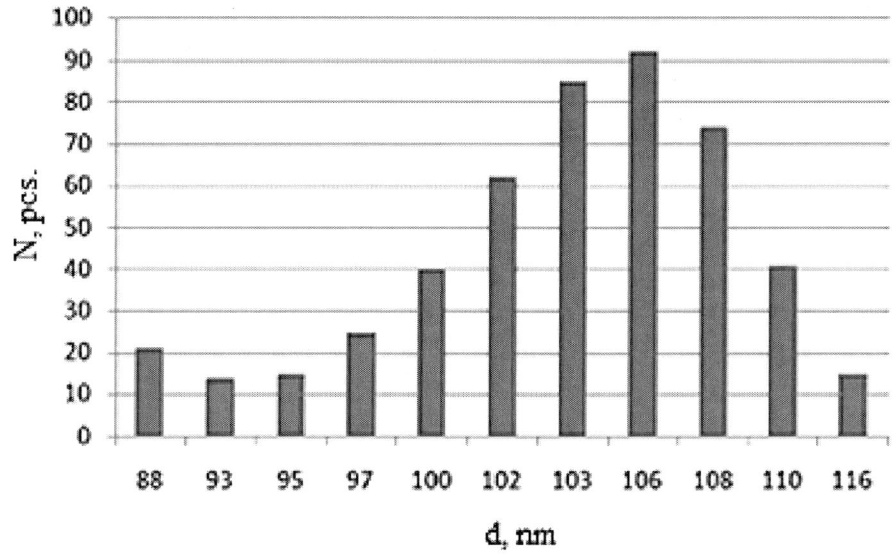

Figure 5. Figure size distribution of pore sizes in the samples produced in an electrolyte based on phosphoric acid.

Table 1. Influence of the electrolyte type on the geometric parameters porous layer and λ_c

Acid	Diameter of pore, nm	Thickness of membrane, nm	Concentration of pores, $1/\mu m^2$	λ_c, nm
sulfuric	20-30	12-18	350	210…280
phosphoric	80-100	4-7	50	450…750

On Figure 5(a) it can be seen that the peak $S_2O_3^{2-}$ (1660-1620 cm^{-1}) is most pronounced for the sample prepared without the glycerol in the electrolyte. Glycerol lines in the spectra should be found among alcohols lines, as it's a simple polyol compound. It may be due to the fact that addition of glycerol promotes that the smaller amount of electrolyte anions incorporate into the structure. Peaks of Al_2O_3 (3500-3400 cm^{-1}) are well identified in the spectra, but there are some fluctuations associated with various modifications of Al_2O_3. According to the spectra, samples obtained in an electrolyte based on solutions of H_2SO_4, are characterized by a modification of the mixed between: γ-Al_2O_3 [3500 cm^{-1}] and ι-Al_2O_3 [3450 cm^{-1}] [12]. Samples obtained in an electrolyte based on solutions of H_3PO_4 have a maximum line in the range of 3450–3455 cm^{-1}. This fact indicates the presence of modifications ι-(iota) and shifts to α-Al_2O_3. In other words, membranes with a smaller porous have a structure closer to sapphire.

In the case of highly ordered periodic structure, the optical spectrum contains fluctuations in the interval [2400-1700 cm^{-1}]. Analyzing of oscillations and knowing the refractive index of Al_2O_3 allows calculating membrane thickness:

$$nd = \frac{1}{4(C_1 - C_2)},$$

where n is Al_2O_3 refractive index (1,6–1,7), d is membrane's thickness, (C_1-C_2) – half-oscillation length (≈100 nm). The calculation shows that the membrane thickness is about 15 μm, what was confirmed by the SEM images.

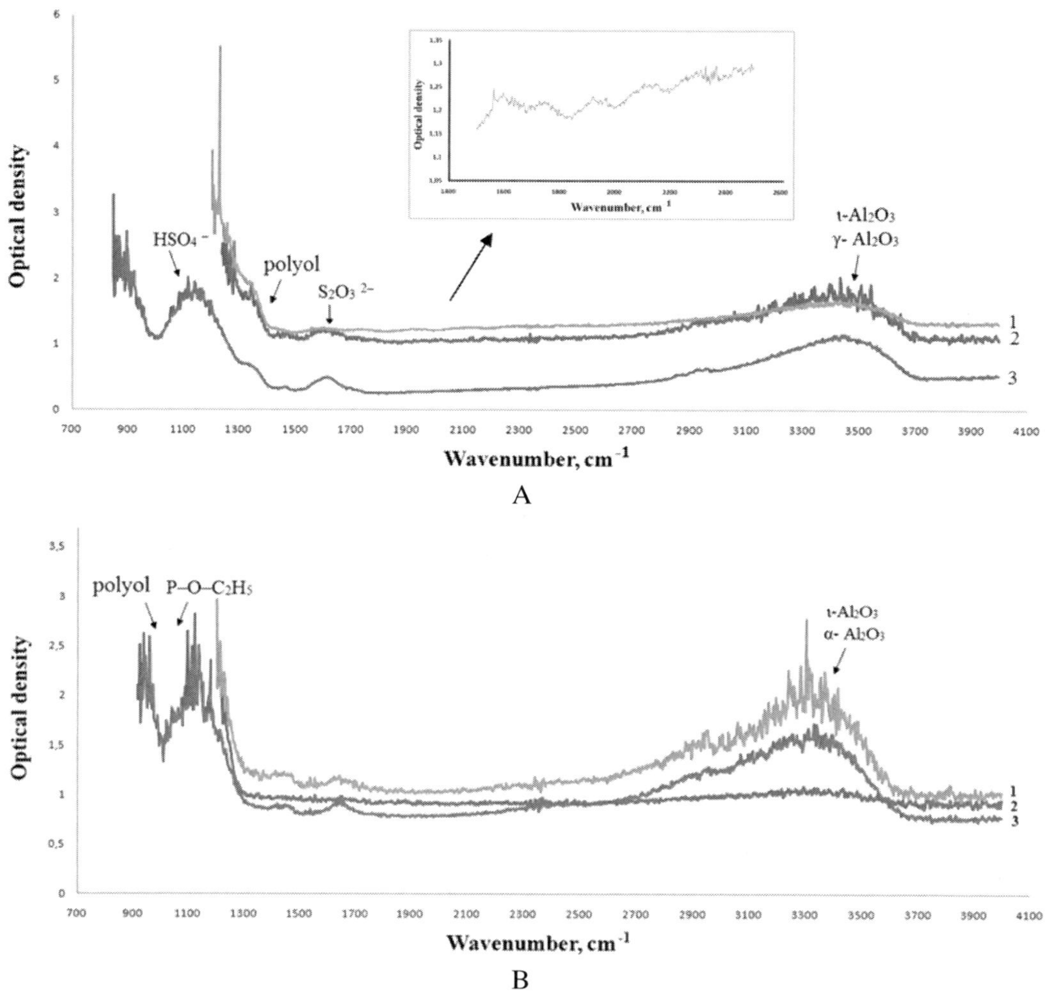

Figure 6. A & B: Optical density spectrum for membranes obtained in an electrolyte based on solutions of: A) H_2SO_4 (1- with 15% glycerol (highly ordered periodic structure), 2 - with 15% glycerol, 3 – without glycerol) and B) H_3PO_4 (1- with 15% glycerol at T=7°C, 2 - with 15% glycerol at T=0°C, 3 – with 10% glycerol at T=7°C).

Figure 7 presents the combined transmission spectra of a porous alumina membrane in a wide optical range. There are several major areas that allow to give a full analysis of the porous membrane, namely to determine the average size and the dispersion of pore diameter, modification of alumina and type of electrolyte anions embedded into the porous structure, and also to calculate membrane's thickness.

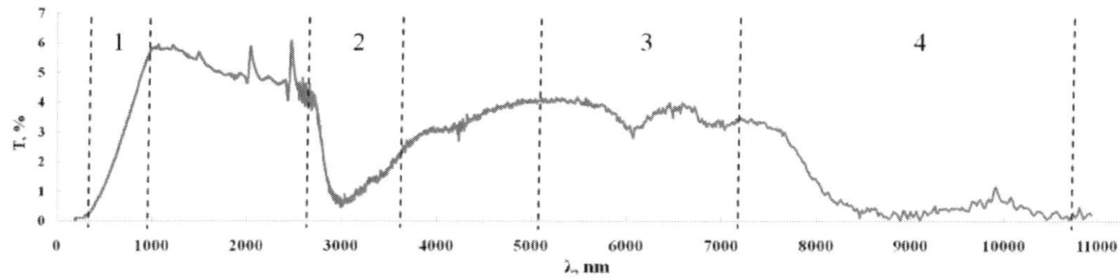

Figure 7. Transmission spectrum in wide wavelength range. Different areas of spectrum are characterize parameters of membranes: 1 – the average size and the dispersion of the pore diameter, 2 – modifications of Al_2O_3, 3 – **membrane's thickness,** 4 – anions of the acid used and the concentration of glycerol in the electrolyte.

Conclusion

Investigations have shown that every porous alumina membrane starts to rise only at a certain critical wavelength, which depends on the electrolyte type in the presence of which membrane was produced. Thus, it is possible to determine the average size and the dispersion of the pore diameter by UV-visible spectroscopy. Also it was established, that IR- spectrums have peaks that characterize the porous membrane from the viewpoint of its structure (modification of the alumina, composition anions embedded into the porous structure, the thickness of the membrane). It is found that membranes with a smaller porous have a structure closer to sapphire. Analyzing oscillations and knowing the refractive index of Al_2O_3 allows calculating membrane thickness. Described method can be used as non-destructive express technique for the determination of pore size in por-Al_2O_3.

The reported study was supported by RFBR, research project No. 14-08-31609 mol_a.

References

[1] Eftekhari A. 2008 *Nanostructured Materials in Electrochemistry*. WILEY-VCH Verlag GmbH and Co. KGaA. Weinheim, (2008) 436.

[2] Afanas'ev A.V., Il'in, V.A., Moshnikov V.A., Sokolova E.N. and Spivak Yu.M. Synthesis of nanoporous and microporous structures by the electrochemical methods. *Biotekhnosfera,* 1–2, (2011) 20-26.

[3] Zhan W., Huangfu Y., Ding G., Yea H. Restrictions of Si-based Genanodots from porous alumina membranes. *Superlattices and Microstructures*, 60 (2013) 73–82.

[4] Macias G., Hernández-Eguía L.P., Ferré-Borrull J. Gold-Coated Ordered Nanoporous Anodic Alumina Bilayers for Future Label-Free Interferometric Biosensors ACS. *Applied Materials & Interfaces*, 5 (16), (2013) 8093–8098.

[5] Mozalev A., Calavia R., Vázquez R.M. MEMS-microhotplate-based hydrogen gas sensor utilizing the nanostructured porous-anodic-alumina-supported WO3 active layer. *International Journal of Hydrogen Energy,* 38(19), (2013) 8011–8021.

[6] Zimina T.M., Muratova E.N., Spivak Yu.M., et al. Technologies of formation and application of nano-layers and nano-porous Al2O3 compositions for micro-and nanotech. *Nano and Microsystem Technique*, 12 (2012) 15-24.

[7] Akiyama T., Ishikawa Y., Hara K. Xylene sensor using double-layered thin film and Ni-deposited porous alumina. Sensors and Actuators B: Chemical. 181 (2013) 348–352.

[8] Foss C.A., Hornyak G.L., Stocketr J.A., et al. Optically Transparent Nanometal Composite Membranes. *Advanced Materials*, 5 no 2, (1993) 135-136.

[9] Shi Z.C., Fan R.H., Zhang Z.D., Qian L. et al. Random Composites of Nickel Networks Supported by Porous Alumina Toward Double Negative Materials. *Advanced Materials*. 24 (2012) 2349–2352.

[10] Choi J, Luo Y, Wehrspohn R B et al. Perfect two-dimensional porous alumina photonic crystals with duplex oxide layers. *Journal of Applied Physics*, 94 (2003) 4757.

[11] Muratova E.N., Spivak Yu.M., Moshnikov V.A., Petrov D.V., Shemukhin A.A., Shimanova V.V. Influence of technological parameters of nanoporous Al2O3 layers' preparation on their structural characteristics. *Glass Physics and Chemistry*, 39(3), (2013) 320-328.

[12] Tolstoy V.P., Chernyshova I.V., Skryshevsky V.A. *Handbook of infrared spectroscopy of ultrathin films*. Published by John Wiley & Sons, Inc., Hoboken, New Jersey, (2003) 710.

In: Nanoscale-Arranged Systems for Nanotechnology
Editors: Kirill L. Levine
ISBN: 978-1-63482-353-1

Chapter 9

INFLUENCE OF NANODIMENSIONAL EFFECTS ON ELECTRIC ADHESION IN ANODIC BONDING MANUFACTURING OF COMPOSITES SEALS

N. S. Pshchelko[1*], V. A. Moshnikov[2] and M. P. Sevryugina[1]
[1]University of Mines, Saint-Petersburg, Russia
[2]Saint-Petersburg State Electrotechnical University, Saint-Petersburg, Russia

ABSTRACT

The appearance of great pulling electric fields at anodic bonding process with a conductor surface to dielectric, turns out to be possible due to the interlayer polarization developing in a dielectric under the action of electric voltage. This results in a negative charge accumulation in a layer of small thickness adjacent to anode. Thus, applied electric voltage is distributed not through the whole thickness of a dielectric but, in fact, is applied to a narrow area of the three-dimensional charge adjacent to anode. Arising strong electric field forces the connected materials to be hold to each other. The physical and mathematical models for force characteristics of a non-uniform electric field are developed. Research of a charge intermittence influence on intensity of an electrostatic field has shown a considerable differences on small distances to the charged surface between carried out and traditional ways of calculation.

Keywords: electric field, discrete charge, electric adhesion

INTRODUCTION

Methods and technologies of gluing, sticking and permanent or temporary fixation of electronic components appear in one form or another in almost all areas of their production. The compounds of materials that combine a variety of mechanical and electrical properties

*nikolsp@mail.ru.

are required in one form or another in almost all areas of production. Specifically, the problem of connection of metals and other conductive materials, including semiconductors, with dielectrics is of particular interest. Many of these materials are nonplastic, insoluble in each other and have high melting point. The connection methods (e.g., fusion welding, resistance welding, diffusion welding) can not always overcome the barrier of incompatibility of metals/semiconductors and non-metallic materials and ensure production of high quality compounds. Therefore, the development of theoretical concepts of adhesion, as well as the development of new technological methods of connection for materials is certainly of interest.

THEORETICAL ASPECTS

Adhesion (from latin *adhaesio* – sticking, coupling, gravity) means a connection between heterogeneous condensed bodies by their contact. Adhesion is manifested in the ability of a contact between (DEL) two heterogeneous bodies to resist contact destruction. Therefore, the phenomena associated with the adhesion are naturally related to surface phenomena, controlled by surface forces. Back in the 40s of the last century, the adhesion theory was proposed, based on the Helmholtz general position of electrical double layer at the interface [1]. Its further development was the electronic theory of adhesion, which investigated the processes of the double layer formation at the bonding between the connected surfaces and its contribution to the strength of the latter [2]. In accordance with theoretical studies, experimental work was developed in which the basic provisions of the electronic theory of adhesion were confirmed and the new area of research – electric adhesion phenomena – was developed. This term refers to the electrical phenomena associated with the establishing and breaking of connections between bodies. The link between electric adhesion and phenomena of triboelectricity and triboluminescence was identified in the experiments, when the emission of fast electrons in adhesive contact violation in a vacuum was detected.

An increase of the surface adhesion is the main task to be solved to obtain a reliable connection between two surfaces. There are two major factors that can increase the adhesion:

1. Diffusion. Strengthening of the contact between two heterogeneous bodies may be provided by diffusion of atoms and molecules of the both, resulting in a blurring of the interface. Sharp interface is transformed into a transition layer whose composition gradually changes from layer to layer. This process stimulated by the external pressure and an increased temperature is the basis for the well-known diffusion welding method.

2. Formation of electrical double layer on the contact. The charge density of this layer is high in the case of a contact between heterogeneous bodies, such as metal and dielectric, when diffusion coalescence is constrained. The role of this factor in the adhesion phenomena, in our opinion, is often underestimated. Indeed, as was shown in [3], an increase in the electric charge density at the interface by the external voltage allows to get the pressure caused by electrostatic forces at the level used in the diffusion welding ($\approx 10^7$ Pa) and to achieve not only temporary, but permanent connection of details. This connection method is called electrical adhesion and is actually the subject of this paper. The role of the DEL in the interface can not be reduced only to the establishment of electrostatic (ponderomotive) forces. e.g., [2] indicates the ability of the electric double layer to neutralize embryonic cracks, formed in the plane of adhesive contact and subject to expansion under external influence (forces, thermal and shrinkage stresses etc.). It is explained by the fact that the force of attraction between oppositely charged plates of the double layer decreases very slowly, rather as a function of time than of distance, differing in this respect from radical molecular forces. So after a brief power influence embryonic cracks can interlock under the influence of electrostatic pressure forces. Similarly, the double layer should increase the fatigue strength of adhesive contacts at periodic influences.

PRACTICAL FINDINGS

The present work is devoted to assessing the influence of discreteness of electric charge distribution in the electric double layer on the characteristics of the electric field and force characteristics of the electric adhesion contact. The formula of a plane capacitor is usually used to calculate the interaction force caused by the attraction of opposite charges, distributed on the planes of the electric double layer:

$$p = \frac{\sigma^2}{2\varepsilon_0\varepsilon}, \tag{1}$$

where p is ponderomotive pressure, ε_0 - dielectric constant, ε - relative dielectric permeability of the material between interacting layers, σ - DEL surface charge density. This implies that the charges on planes are distributed continuously and uniformly. In a stricter approach, it becomes obvious that such model does not correspond to the facts, because electric charge is discrete (the elementary charge e = 1,6·10^{-19} C) and, therefore, the electric field of point charges system is not homogeneous. At the same time, the model for the calculation of the ponderomotive effect of the electric field on the basis of (1) is acceptable in the case when the distance between the planes of the electric double layer is much greater than the distances between the discretely spaced elementary charges. In the case when the distance between planes of DEL is commensurate with the distance between the discretely spaced elementary charges, e.g., during the preparation of electric adhesion compounds [3], the heterogeneity of the electric field manifests itself. Thus, the results obtained by (1) will be inaccurate.

Let us find out, at what distance between planes the formula (1) will not adequately reflect the real physical situation. As an example let us consider a uniformly charged plane with surface charge density σ = 10^{-4} C. As it is known, the electric field created in a vacuum (ε= 1) will be equal to $E_c = \frac{\sigma}{2\varepsilon_0\varepsilon}$ = 5,65·10^6 V / m. The amount of elementary charges per 1 m^2 area that obviously corresponds to the specified value σ is $N = \frac{\sigma}{e}$ = 6,25·10^{14}. Consequently, the average distance between them is $a = \frac{1}{\sqrt{N}}$ = 4·10^{-8} m. Thus, at the distance between the electric double layer planes corresponding to the order of tenths of a micron, the formula (1) can give untrue value. Let us note that it is the afore mentioned distance between the charges that is used in the preparation of electric adhesion compounds, since the clearance between the connected parts is caused mainly by roughness of the connected surfaces that are usually treated like a 11th class purity. Moreover, in the process of obtaining electric adhesion compounds surfaces are closed together and the distance between them decreases. For reference, we note that the value of the arithmetic deviation of the rough surface profile for items treated by the 11th class is 0.08 μm, and for treated by 14th class - 0.01 μm.

Thus, at small distances between the charged planes the discreteness of electric charge distribution should be taken into consideration. Indeed, in experiments [2] ponderomotive forces are often observed that have a much bigger value than calculated according to (1). One of the reasons for this is probably a discrete charge distribution.

Let us consider the elementary charges localized in a layer consisting of several planes (Fig. 1). The thickness of a layer with spatially distributed charge is $l \cdot a$, where l is the number of planes, and a is the minimum distance between the charges in the system. We specify the same total charge for this layer volume as in the case of a continuous distribution of charge on the plane. We define the electric field $\vec{E}$ at a point M, created by a system of discrete spaced elementary charges on a distance a. In our case, $\vec{E}$ is a field of a charged layer with linear dimensions on the plane xOy (model coordinate system - Figure 1), that exceed the distance from the current point in space to the charged plane. It allows to move to a one-dimensional problem in a half-space bounded by an infinitely extended charged plane.

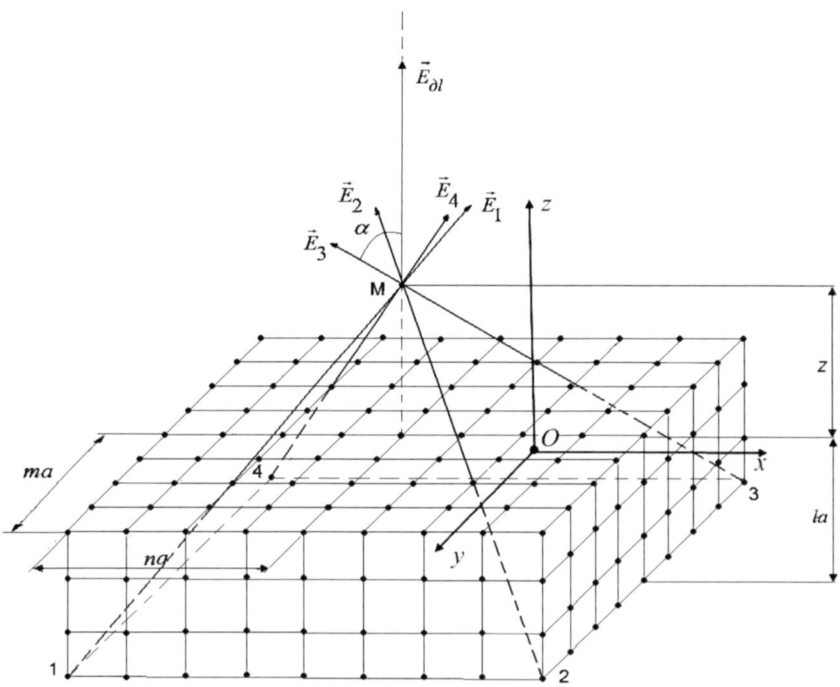

Figure 1. Layer model with intermittent charge distribution. Explanatorily depicts several vectors and their resultant tension at a point located on one normal with the charge. n, l and m are integers.

We consider two situations - with a continuously distributed charge and a localized one due to the existence of an elementary charge ($q_1 = e = 1{,}6{\cdot}10^{-19}$ C) - and estimate the difference in the results of the calculation of the electric field intensity. Basing on the symmetry of the model, the principle of superposition, the geometric considerations and the Coulomb law, the cumulative effect of charges shown in Figure 1 can be reduced as a result of the sum:

We consider two situations - with a continuously distributed charge and a localized one due to the existence of an elementary charge ($q_1 = e = 1{,}6{\cdot}10^{-19}$ C) - and estimate the difference in the results of the calculation of the electric field intensity. Basing on the symmetry of the model, the principle of superposition, the geometric considerations and the Coulomb law, the cumulative effect of charges shown in Figure 1 can be reduced as a result of the sum:

$$\mathring{A}_{\partial} = \sum_{l=0}^{L-1}\left(\sum_{n=0}^{\infty}\sum_{m=1}^{\infty}\frac{q_1}{4\pi\varepsilon_0\varepsilon}\frac{4(z+al)}{\left((z+al)^2+a^2(m^2+n^2)\right)^{3/2}}+\frac{q_1}{4\pi\varepsilon_0\varepsilon(z+al)^2}\right). \quad (2)$$

Numerical calculations show the insignificance of the influence that the layer thickness has on the resulting electric field. Therefore, it is enough to consider the case L=1 and therefore for the electric field plane with a discrete charge distribution we obtain the expression:

$$\mathring{A}_{\partial} = \left(\frac{q_1 z}{\pi\varepsilon_0\varepsilon a^3}\sum_{m=1}^{\infty}\sum_{n=0}^{\infty}\left(\frac{z^2}{a^2}+m^2+n^2\right)^{-3/2}\right)+\frac{q_1}{4\pi\varepsilon_0\varepsilon z^2}; \quad (3)$$

Direct use of (3) leads to computational problems, so for practical use of the expression obtained it must be approximated with an analytical formula. Our computations (due to their large size, they are not given) show that in most practical cases, you can use the expression:

$$\mathring{A}_{\partial} \approx \frac{\sigma}{\pi\varepsilon_0\varepsilon}\left[arctg\left(\frac{z}{a}\right)+\frac{\pi a}{2z}\right]. \quad (4)$$

We show that the expression above accurately reflects the situation in the limit case. Let us consider a field at $a=\sqrt{\frac{q_1}{\sigma}} \to 0$ that corresponds to the model of a continuous charge distribution. The argument $\frac{z}{a} \to \infty$, consequently $arctg\left(\frac{z}{a}\right) \to \frac{\pi}{2}$, at the same time function $\frac{\pi a}{2z} \to 0$, proportional to an infinitesimal value, can be ignored. We obtain

$\mathring{A}_{\partial} \rightarrow \mathring{A}_{l}$ that corresponds to the classical case of a continuous charge distribution on the plane. Thus, we can define the limits of applicability of classical formulas for the calculation of dependence of electrostatic field intensity on the distance to the charged object, Figure 2.

Calculations show that the consideration of charge discreteness leads to significant amendments in the direction of increased values of field intensity and ponderomotive force, acting on the charge at a relatively small distance from the heterogeneously charged plane.

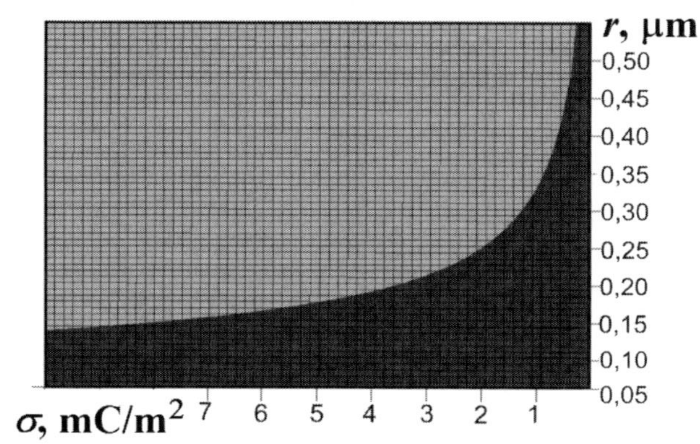

Figure 2. The limits of applicability of the classical formula for the calculation of the electric field intensity of a charged plane in case of the charge distribution in the layer of small finite thickness. The bright field represents the area, where relative error $\Delta E_{rel} = \frac{E_{\partial} - E_c}{E_c} \cdot 100\%$ of classical formula application does not exceed 10%, the dark field - the area where it does exceed 10%.

To appreciate the differences between values E_{∂} и E_c one can also use the data on the calculated values of the ponderomotive pressure $p = \frac{1}{2}\varepsilon_0 \varepsilon E^2$, shown in Figure 3.

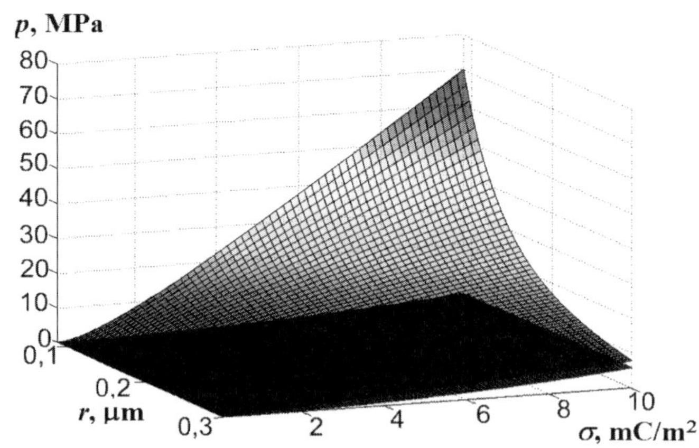

Figure 3. Dependence of the ponderomotive pressure on the distance from the plane and from the surface charge density. The dark surface corresponds to a continuous distribution, the bright one - to a discrete distribution of the same amount of charge on the plane.

CONCLUSION

Thus, at small distances to the charged plane electric field strength *E* by magnitude is significantly higher than the one calculated by the classical formulas. Considering that the field is heterogeneous, due to the gradient forces, it may cause attraction even electrically neutral objects, e.g., atoms sprayed on the substrate [5] Different aspects of small interactions are studied in [4]. Mathematical equations provided in this work are believed to be sufficient for calculating these effects.

REFERENCES

[1] Damascene B.B., Petriy O.A. *Introduction to electrochemical kinetics*. Moscow, (1983), 400.

[2] Derjagin B.V., Krotova N.A., Smilga V.P. *Adgesion of solid bodies*. Moscow, Nauka, (1973), 280.

[3] Nagorny V.S., Pshchelko N.S., Sidorova N.P. Analysis of dynamics of formation process of anodic bonding contact//*Scientific and technical sheets of St.-Petersburg polytechnical university*. 6, (2009),166-170.

[4] Pshchelko N.S. Use of electric field for increase of adhesion of electrospending films to dielectric substrates at vacuum deposition // *Vacuum technics and technology*, 1, (2010) 31-36.

[5] Afanas'ev V. P, Chigirev D.A., Pshchelko N.S., Sidorova N.P. Influence of constant electric field on processes of sedimentation of thin metal films of platinum by a method of ionic-plasma dispersion // *News of HIGH SCHOOLS of Russia. Radio electronics*. 6, (2010), 59-65.

In: Nanoscale-Arranged Systems for Nanotechnology
Editors: Kirill L. Levine
ISBN: 978-1-63482-353-1

Chapter 10

CORRELATION NATURE OF PHASE TRANSFORMATIONS IN NANOCOMPOSITES ON THE BASIS OF VO_2

***A. V. Ilinskiy*[1]*, O. E. Kvashenkina*[2] *and E. B. Shadrin**[*,1]**
[1]Ioffe Physical Technical Institute of the Russian Academy of Sciences, Russia
[2]St. Petersburg State Polytechnical University, Russia

ABSTRACT

A complex concept concerning the multistage process of thermal semiconductor–metal phase transition in vanadium dioxide is discussed. According to the concept, the phase transformation proceeds in three stages: two stages occurring in a system of strongly correlated electrons and one stage in lattice system which increases the crystal lattice symmetry by phase transformation. This concept generates a possibility of creation of a wide set of devices with new applications.

Keywords: vanadium dioxide, phase transition, strong correlations

INTRODUCTION

In the last decade, interest in studies of the semiconductor–metal phase transition (PT) in vanadium dioxide has been renewed. On one hand, vanadium dioxide presents a model of highly correlated systems and, on the other hand, it is a promising material for various technological applications. In the series of metal oxides, vanadium dioxide occupies a special place, since vanadium is an element with an unfilled *d* shell and, therefore, demonstrates profound correlation effects. Due to electron correlations, vanadium dioxide exhibits unique properties defined by a large variety of types of chemical bonds between oxygen and vanadium atoms in the compound, including bonds involving *d* electrons. Vanadium dioxide

[*]Corresponding author: Prof. E.B. Shadrin, Professor of Ioffe Institute, St. Petersburg, Russia. Tel.: + 79095798026. E-mail: shadr.solid@mail.ioffe.ru.

is a material that exhibits first order PT from the monoclinic (C_{3v}) phase to the tetragonal (D_{4h}) phase (the semiconductor–metal phase transition). In VO_2 single crystals, this transition is observed at the temperature T_c = 67°C (340 K) [1]. Upon such transition, both the optical and electrical parameters of the material undergo a stepwise change. Specifically, the refractive index changes from 2.5 in the monoclinic phase to 2.0 in the tetragonal phase, and the change in conductivity at the PT in single crystals is ~10^5. At temperatures higher than T_c, the material possesses metallic conductivity, with the concentration of charge carriers ~10^{22} cm^{-3}, and the tetragonal symmetry of a rutile-like crystal lattice.

Crystal Structure of Vanadium Dioxide

As it is known from X-ray structural analysis [1], the unit cell of the lattice of the tetragonal VO_2 phase contains two vanadium atoms and four oxygen atoms (Figure 1).

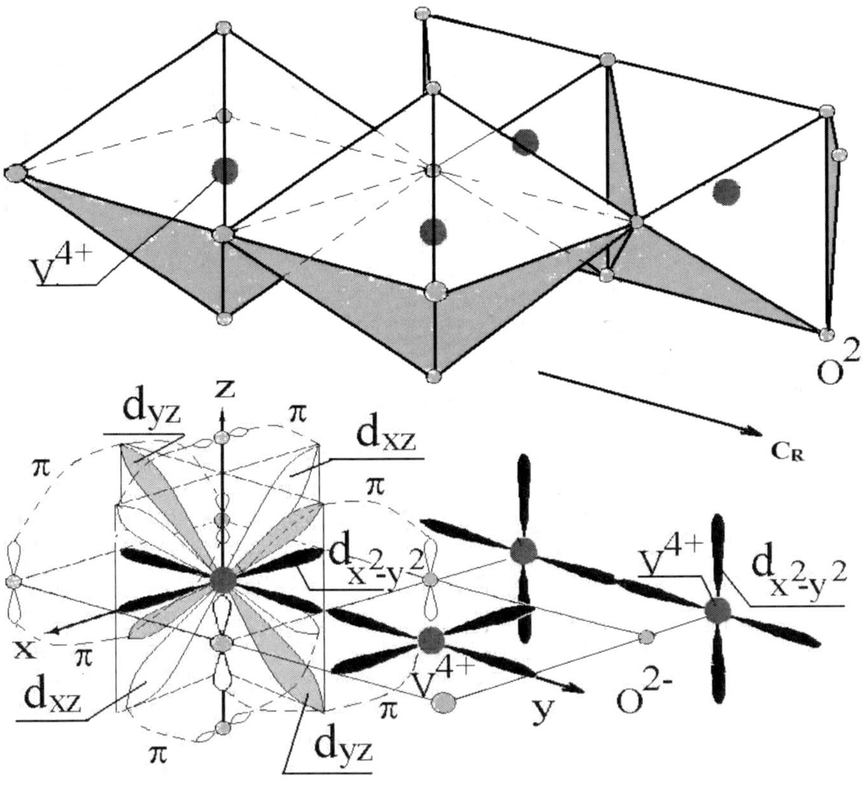

Figure 1. Schematic representation of the 3*d* orbitals of a V^{4+} ion in an oxygen octahedron.

The valence bonding theory says that, in such a unit cell, due to $3d^24s^14p^3$ hybridization, each vanadium atom forms six σ-bonds with six oxygen atoms, and each of them has three sp^2 hybridized orbitals. All vanadium atoms are at the centers of oxygen octahedrons. The remaining three nonbonding 3*d* orbitals (d_{xz}, d_{yz}, d_{x2-y2}) of the vanadium atom, with one electron belonging to them, do not participate in the formation of the skeleton of σ-bonds of the oxygen octahedron. Except for σ-bonds, in the oxygen octahedron, there are saturated π-bonds generated from orbitals of vanadium (V^{4+}) ions and $2p_z$ orbitals of oxygen ions. According to the theory of molecular orbitals, these π-bonds form π and π* electron energy levels. In the crystal, these levels broaden into π and π* energy bands. The d_{x2-y2}-orbital has zero overlapping with all of the ligand orbitals. It is nonbonding and has the lowest energy among the orbitals that do not have connections with the V^{4+} ion. The d_{x2-y2}-orbital is not involved in the formation of the octahedron and takes a one electron remaining to the all d_{xz}, d_{yz} and d_{x2-y2} orbitals. According to X-ray structural data [1], the oxygen octahedrons are arranged so that the tetragonal phase contains quasi one dimensional (1D) chains of vanadium ions located at the centers of oxygen octahedrons; the chains are parallel to the rutile C_R axis (Figure 1). Thus the orbital containing one electron for each ion with branches arranged along the x and y axes can be responsible for a nonzero overlap integral only with a similar orbital of V-ions in neighboring octahedrons arranged along the C_R axis. Thus, vanadium dioxide demonstrates a unique case, in which a 1D chain of hydrogen-like atoms, each with one electron at the d-orbital, is stabilized in space mainly by σ-bonds of oxygen octahedrons. However, in contrast to the electron in a hydrogen atom, this single electron is located, and on the d-orbital but not on the s-orbital. Unlike the spherically symmetric s-orbital, the d-orbital has cross-like branches, which can form 1D metal-type bonds with the orbitals of similar atoms of neighboring octahedrons arranged along the C_R axis.

SAMPLES

Our experiments were based on the analysis of the optical and electrical temperature hysteresis loops of mentioned phase transition. For optical measurements, we synthesized on a 0.8 -1 mm-thick and 20x20 mm area silicon substrate an interference structures comprising a high reflectivity aluminum mirror with 30–60 nm thickness, deposited directly on the substrate and having on itself a 70–100 nm-thick VO_2-film. The second mirror of the interferometer was deposited on the film. The interference structure was synthesized according to the standard technique. Such design of the interferometer was necessary therefore that the jump of the reflectivity in VO_2 films during a PT is small and amounts to only a few percentage points [1]. For this reason, it is convenient to study optical hysteresis loops by the method of interference enhancement of the contrast, in which a VO_2 film is placed between the mirrors of a Fabry–Perot interferometer. As a result, the jump in the reflectivity during the PT is not less than 50%, due to the shift of the interference minimum of the interferometer reflectivity towards shorter waves. This is due to the fact that a PT gives rise to jumps in the optical constants (the refractive index at $\lambda = 633$ nm changes from 2.5 to 2.0, while the extinction changes from 0.4 to 0.6 in the semiconducting and metallic phases, respectively) as well as in the resistivity of the material (from 10^{-6} to 10^{-1} W/m).

Samples of VO_2 photonic crystals were synthesized using artificial opals with a face-centered cubic lattice formed by close-packed monodisperse spheres of amorphous SiO_2 as the template. About 26% of the total volume of the opal template is accounted for by interconnected voids that are accessible to filling by other substances. The average sphere diameter was 230±5 nm. The opals had a polydomain structure, the size of a single domain with a high degree of ordering of SiO_2 spheres varying within 30–100 nm. The photonic crystal was formed on the basis of VO_2 by two stages. In the first stage, voids of the opal template were preliminarily filled with a solution of vanadium pentaoxide in nitric acid and then V_2O_5 was reduced to VO_2 through the high-temperature annealing of a sample in a vacuum. In the second stage, the composite was subjected [2] to inversion by etching out the skeleton of SiO_2 spheres with hydrofluoric acid.

By present time are in detail investigated X-ray, morphological, optical, electrical and tension properties of pure and doped VO_2-nanocomposites [1-19]. Namely, in papers [2, 12] are described results of investigations of VO_2-opal composites, in papers [9-12] - of superfast (femtosecond) optical and X-ray response of VO_2 crystals, in papers [15] - of hydrogenated VO_2-films, in papers [18,19] - of thermoreflectance and Raman properties of VO_2-interferometer, The detailed analysis of these results has allowed us to formulate the new concept of the mechanism of phase transformations in vanadium dioxide.

Global Mechanism of Phase Transformations in Vanadium Dioxide

For a chain of single electron atoms, the Schroedinger equation with the Hamiltonian that takes into account the interaction between only the nearest neighbors has an exact solution and can be written as [3]

$$[d^2\Psi(y)/dy^2] + (2m/\hbar^2)[E-V(y)]\Psi(y)=0,$$

Here, V(y) is the potential energy equal to the sum of the potential energies of an electron in the field of individual ions, without regard for a small addition, specifically, for the interaction energy of the electron of this ion with his neighboring ion (the direction of the y axis coincides with the direction of the rutile C_R axis (Figure 1)). The solution of the Schroedinger equation predicts the existence of allowed and forbidden energy bands for free $3d^1$ electrons belonging to vanadium ions, which are members of the 1D chain. The electrons of $3d$ orbitals occupy the lowest states of the allowed $3d\|$ band, and according to the Pauli principle, this band is half filled with pairs of electrons with antiparallel spins. These electrons occupy the lower half of the band, since the number of levels in the band is equal to the number of members in the chain. The upper half of the allowed band remains unfilled, as happens in a typical metal; i.e., having one electron, the nonbonding orbitals of vanadium ions form a system of metal type bonds with neighboring ions of the chain. A specific feature of this system of bonds is that each atom supplies only one electron to the standard metal bond. In such a first approximation, metal-like conductivity along the 1D chain appears due to the half filling of the allowed electron band and is bound to persist at all temperatures, including low temperatures. Thus, such a system is prevented from undergoing any phase transformations.

Abandonment of the Rigid Chain Concept

In according to X-ray structural data [1], at $T < T_c$, the lattice symmetry lowers from the tetragonal to monoclinic type, since neighboring atoms in the chain approach in pairs to form stable ion pairs (dimers). Since there are two vanadium atoms in the VO_2 unit cell, the number of pairs is equal to the number of unit cells. The dimers appearing in chains distorts each unit cell and, as a result, distorts the entire crystal lattice, with the above-mentioned lowering of the lattice symmetry. Pairing of vanadium ions means that the spacing between vanadium ions within a pair (within a dimer) is smaller than the spacing between the pairs themselves; therefore, the lattice period along the chain doubles in comparison with the initial period. Because of such structural transformation, i.e., the Peierls transition, a band gap is formed in the electron spectrum [3], and the vanadium dioxide transforms into a semiconductor. The calculated band gap of such a semiconductor lies in the range 0.2–0.5 eV. In this case, the $3d\|$ band breaks into two subbands, each containing half of the total number of levels in the initial $3d\|$ band. According to the experiment [4], the splitting between the subbands is 2.5 eV, i.e., so large that we have to suppose that the π^* band is located between $d_{x2\text{-}y2}$ – subbands, at an energy of 0.7 eV above the lower $3d\|$ subband. With the band gap of 2.5 eV, the lower $3d\|$ subband is almost completely filled with electrons, whereas the upper subband is practically empty. Therefore, in the model of a not rigid chain, the VO_2 crystal is a semiconductor with a 0.7 eV band gap between the lower $3d\|$ subband and the π^* band (Figure 2), as reliably established experimentally.

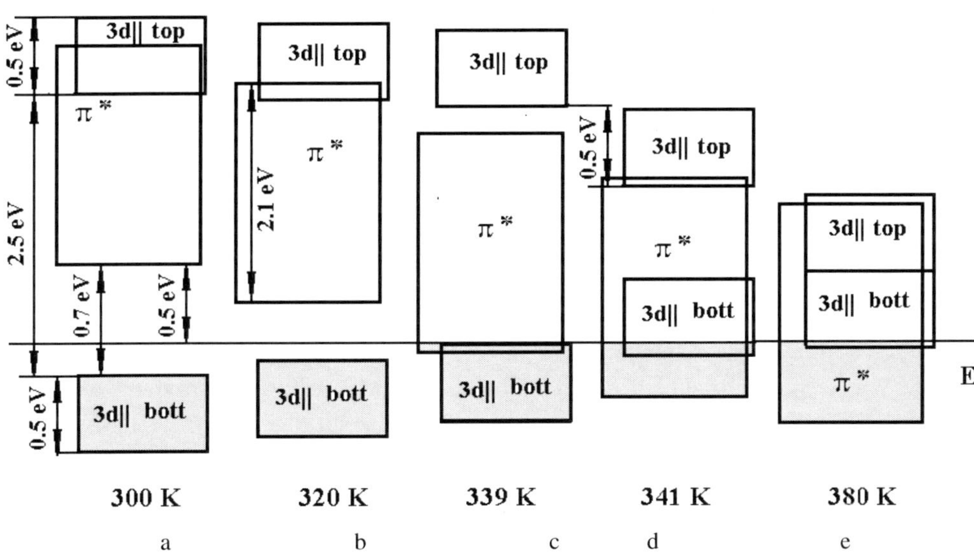

Figure 2. Energy band diagram of VO_2: (a,b) the low temperature monoclinic semiconductor phase (300, 320 K); (c) the monoclinic phase near the structural transition temperature (339 K); (d) the tetragonal metal phase near the structural transition temperature (341 K); and (e) the tetragonal metal phase at high temperature (380 K).

The scheme modified in this direction can be improved further by taking into consideration the possibility of the thermal ionization of dimers, which allows vanadium dioxide to not remain a semiconductor at temperatures close to the PT temperature. In fact, at

$T > 67°C$, vanadium dioxide is a metal, as follows from a large body of experimental data [1]. All of these data suggest that the lattice increases its symmetry from the monoclinic to the tetragonal type at $T > T_c$. However, simple estimates show that, with the 0. 7 eV band gap between the almost completely filled $3d\|$ subband and the practically empty π^* band, increasing temperature to $T_c = 67°C$ cannot induce the complete destruction of dimers via the thermally activated transfer of electrons into the π^* band.

In addition, although the model for the doubling of the period of a 1D chain along the C_R axis (the Peierls model) allows an exact solution for VO_2, the model gives a value for the splitting between the two $3d\|$ subbands of no larger than 0.5 eV, whereas the splitting observed experimentally is 2.5 eV [4].

Consideration of Correlation Effects

Further development of the model of PT processes is based on the well-known Mott idea [5], consisting of the following: to solve the problem concerning the behavior of an electron in the field of a 1D chain of vanadium ions, the one particle approximation is inadequate, and it is essential to take into account the rather strong correlation interaction between electrons themselves. This interaction leads to dependence of the energy position of the bands on their occupation by electrons [6] and creates a correlation addition of ~2.0 eV to gap between the $3d\|$ and $3d\|$ subbands. In fact, even in the context of a simple model, in which a V–V dimer is represented as a hydrogen molecule, consideration for correlation interaction between electrons in the molecule yields the theoretical $3d\|$–$3d\|$ splitting at ~2.5 eV. Thus, we obtain an additional increase in the splitting of the $3d\|$ band to ~2.5 eV due to correlation interactions; i.e., we are dealing with the formation of a wide gap semiconductor, whose band gap consists of the Mott gap of 2.0 eV and the Peierls gap of 0.5 eV. With this result, we finish the construction of a qualitative, but experimentally justified model of the band structure of the material's ground state. Obtaining a rigorous theoretical solution to a multiparticle problem is a rather complex challenge in itself. However, recent computer calculations in the context of the G0W0 + COHSEX approximation [6] show that, in the vicinity of the Fermi level *EF* in the energy spectrum of $3d$ electrons of VO2, a 0.7 eV band gap is formed and the splitting of the $3d\|$ band is 2.5 eV. The same calculations show that the energy position of the bands depends on the degree of band filling with electrons, so that, as electrons are transferred to the unfilled bands, the energy gaps become narrower.

Thus, according to the multistage scheme developed specifically for the PT in vanadium dioxide, upon heating of the samples, correlation effects induce a decrease in the energy gap between the lower $3d\|$ subband and the π^* band from the initial band gap of 0.7 eV, since the bands move in correlation with the thermally activated transfer of electrons from the lower $3d\|$ subband to the π^* band (Figures 2a, 2b). In this case, positive feedback between the band gap and the band populations is produced: a decrease in the band gap promotes an increase in the rate of the thermal generation of electrons into the π^* band, and in turn, this increases narrows the band gap. It should be noted that, for the same reason, the upper $3d\|$ subband falls, but more slightly.

But simple estimations show that the thermal "tails" of Fermi distribution of electrons between energy levels are short (~0.030 eV) and are unable to provoke thermal decomposition of the dimers that is requisite for the structural PT. With this circumstance taken into consideration, we think that the mechanism for the semiconductor–metal PT is such as described below. With increasing temperature, correlation interaction between electrons narrows the energy gap between the bottom of the π^* band and the top of the $3d\|$ subband and they touch the Fermi level (Fig. 2c). This process initially proceeds without any noticeable breakage of the dimers. As the sample is heated further, the bottom of the π^* band lowers because correlations and do below the Fermi level, and as result, the transitions of electrons from the lower $3d\|$ subband to the π^* band are intensified; thus, the destruction of dimmers becomes more efficient. However, the complete decomposition of dimers, which is necessary for the structural transition to occur, only takes place, if the vast majority of electrons from the $3d\|$ subband forming the dimers transit to the π^* band. In turn, this is only possible if the number of states in the π^* band is substantially larger than that in the lower $3d\|$ subband, which in practice is indeed in the case for VO_2, since the number of levels in the lower $3d\|$ subband and the π^* band are, correspondingly, $N/2$ and $2N$, where N is the number of vanadium atoms. After the above-mentioned interaction, the $3d\|$–π^* transitions are intensified in a narrow temperature region below the structural PT to the metal phase. Correlation interaction not only lowers the energy position of bands during their filling with electrons, but also extends "the tails" for electron distribution among the energy levels, thus reducing the step in the distribution function at the Fermi level (this step is commonly referred to as the Migdal jump [3]). The combined effect of the intensification of electron processes and the correlation-induced extension of the tails greatly facilitates the disruption of dimers by efficiently pumping electrons out from their σ bonds, which belong to the lower $3d\|$ subband. As the critical concentration of broken dimers is attained, the structural PT to the tetragonal phase proceeds abruptly. Thus, in this complex model of the PT, the Mott electron transition, at which the conductivity of the material gradually increases as a result of the correlation induced narrowing of the band gap, is primary and, upon heating of the sample, not only initiates the gradual approach of the material to the region of stimulated disruption of σ bonds in dimers, but also ensures the process of this accelerated breakage as such. It should be noted that, in itself, the onset of the accelerated breakage of dimers does not correspond to the instant of structural PT occurrence. Only at the instant when the critical concentration of broken dimers at the temperature $T = T_c + \Delta T$ is attained, a nucleus of the new tetragonal phase located within the old phase extends over the whole VO_2 film nanocrystal at the speed of sound, that reflecting the martensitic character of the PT in VO_2. (Here, ΔT is the deviation of temperature from T_c; the deviation is necessary for a viable overcritical sized martensitic nucleus of the new phase to be formed [7].) Thus, only at the point $T = T_c + \Delta T$ on the temperature scale, the material exhibits a stepwise structural PT to the new tetragonal metal phase under the influence of forces that are produced by σ-bonds of the oxygen octahedrons and vanadium ions that are returning to the centers of these octahedrons (Fig. 2d). It should be noted that, due to the structural PT, the upper $3d\|$ subband unevenly shifts down in energy scale only by the Peierls energy 0.5 eV because of the disappearance of doubling of period of the 1D chain. This means that the material transforms into the metal phase, with conduction occurring only over the π^* band; at the same time, to relation of the $3d\|$ band, the material remains a Mott insulator with a band gap of 2.0 eV. The doublet structure of the $3d\|$ band at this temperature is suggested by experimental data on the photoemission spectra of VO_2 excited by synchrotron radiation with the photon energy 60 eV [4]. The above proposed concept makes the prediction that, due to the sharply increased rate of correlation induced narrowing of the $3d\|$ band gap, the Peierls stepwise lowering of the upper 3*d*|| subband in turn is capable of stimulating the Mott insulator–metal electron transition now within the 3*d*|| band. In other words, the Peierls stepwise lowering of the upper 3*d*|| subband is capable of stimulating the decrease in the energy gap between the upper and lower 3*d*|| subbands from 2.0 eV to zero, as the material is heated further to temperatures above T_c. This is bound to yield extra metallization of the sample (Figure 2e).

CONCLUSION

The described mechanism of phase transformation allows to explain a wide set of experimental results. They are the results of investigation of mode locking of YAG: Nd - laser [8], femtosecond spectroscopy [9-12], results of influence of pressure, hydrogenation and electron beam on a hysteresis of physical parameters of VO_2 - nanobeams, and VO_2 thin films [13-17]. The data on investigation of hysteresis of the photon zone in opal - VO_2 composites [2] can be reasonably interpreted too. Besides the concept of correlation nature of phase transformations in VO_2-nanocomposites finds its confirmation in experiments on temperature dependence of Raman spectra in these materials [18, 19].

In summary we can say, that the results of theoretical and experimental researches of VO_2-nanocomposites received to the present time can be used as the base for development and creation of such practically important devices, as gas sensors, strain sensors, thermo-sensors, vibro-sensors, IR-bolometers, superbroadband transistors and superfast optical limiters.

REFERENCES

[1] W. Bruckner, H. Opperman, W. Reichelt, G. Wolf, F.A. Chudnovsky, E.I. Terukov. *Vanadium dioxide* (Berlin, Academy-Verlag, 1983).

[2] V. G. Golubev, D. A. Kurdyukov, A. B. Pevtsov, A. V. Sel'kin, E. B. Shadrin, A.V. Ilinskiy, R. Boeyink. *Semiconductors,* Vol. 36, No. 9, pp. 1043–1047 (2002).

[3] S. V. Vonsovskii and M. I. Katsnelson, *Quantum Solid State Physics* (Nauka, Moscow, 1983) [in Russian].

[4] S. Shin, S. Suga, M. Taniguchi, M. Fujisawa, H. Kanzaki, A. Fujimori, H. Daimon, Y. Ueda, K. Kosuge, and S. Kachi, *Phys. Rev*. B 41, 4993 (1990).

[5] N. F. Mott, *Metal–Insulator Transitions* (Nauka, Moscow, 1979; Taylor Francis, London, 1974).

[6] H. Jiang, R. I. Gomez Abal, P. Rinke, and M. Scheffler, *Phys. Rev*. B 82, 045108 (2010).

[7] A. L. Roytburd. UFN, 113, 69 (1974).

[8] E. B. Shadrin and A. V. Il'inskii, *Phys. Solid State* 42,1126 (2000).

[9] C. K'ubler, H. Ehrke, R. Huber, R. Lopez, A. Halabica, R. E. Haglung, Jr., and A. Leitenstorfer, *Phys. Rev. Lett.* 99, 116401 (2007).

[10] Hyun_Tak Kim, Yong Wook Lee, Bong Jun Kim, Byung Gyu Chae, Sun Jin Yun, Kwang Yong Kang, Kang Jeon Han, Ki Ju Yee, and Yong Sik Lim, condmat/0608085v1 [condmat. str el], No. 3, 1 (2006).

[11] F. Cilento, C. Giannetti, G. Ferrini, S. D. Conte, T. Sala, G. Coslovich, M. Rini, A. Cavalleri, and F. Parmigiani, *Appl. Phys. Lett.* 96, 021102 (2010).

[12] D. A. Mazurenko, R. Kerst, J. I. Dijkhuis, A. V. Akimov, V. G. Colubev, A. A. Kaplyanskii, D. A. Kurdyukov, and A. B. Pevtsov, *Appl. Phys. Lett.* 86, 041114 (2005).

[13] E. Arcangeletti, L. Baldassarre, D. Di Castro, S. Lupi, L. Malavasi, C. Marini, A. Perucchi, P. Postorino.arXiv:cond-mat/0611281v1 [cond-mat.mtrl-sci] (2006).

[14] V.N. Andreev, V.A. Klimov, *Physics of the Solid State*, Vol. 53, No. 3, pp. 577–582 (2011).

[15] A. V. Ilinskiy, O.E. Kvashenkina, E.B. Shadrin, *Semiconductors,* Vol. 45, No. 9, pp. 1153–1157 (2011).

[16] A. A. Bugaev, A. I. Gavrilyuk, A. A. Gur'yanov, V. P. Zakharchenya, and F. A. Chudnovskii, *Sov. Tech.Phys. Lett.* 4, 27 (1978).

[17] A. V. Il'inskii, V.Yu. Davydov, R.A. Kastro, O.E. Kvashenkina, M.E. Pashkevich, E.B. Shadrin *Technical Physics Letters*, Vol. 39, No. 8, pp. 705–708 (2013).

[18] V. N. Andreev, V. A. Klimov, and M. E. Kompan, *Phys.Solid State* 54 (3), 601 (2012).

[19] I. N. Goncharuk, A.V. Ilinskiy, O.E. Kvashenkina, E.B. Shadrin, *Physics of the Solid State*, Vol. 55, No. 1, pp. 164–174 (2013).
http://arxiv.org/pdf/cond-mat/0608085v1.pdf
http://arxiv.org/pdf/cond-mat/0611281v1.pdf

In: Nanoscale-Arranged Systems for Nanotechnology
Editors: Kirill L. Levine
ISBN: 978-1-63482-353-1

Chapter 11

ACID-BASE ASPECT OF CONTROL OF NANOCOMPOSITE ELECTRICAL PROPERTIES

Maxim Sychov*, Andrey Syrkov[1], Yoichiro Nakanishi[2], Kazuhiko Hara[2], Hiroko Kominami[2] and Hidenori Mimura[2]
St. Petersburg State Institute of Technology (Technical University)
St. Petersburg, Russia
[1]National Mineral Resources University, St. Petersburg, Russia
[2]Research Institute of Electronics, Shizuoka University
Johoku, Naka-ku, Hamamatsu, Japan

ABSTRACT

In this paper it is shown that electrical properties of polymer-inorganic composites (including nanocomposites) depend strongly, among other factors, on the acid-base interactions between polymer and surface of filler solid particle. Several examples are given supporting the introduced approach of controlling properties of composite via interface interactions.

Keywords: acid-basic properties, surface, interface, polymer-inorganic composite

INTRODUCTION

Polymer-inorganic composites are used in electronic devices, printed electronics, embedded electronics applications, etc. Properties of composites, including electric ones, depend on composite structure. Figure 1 shows two possible types of nanoparticles distribution within the composite matrix: statistical and ordered. It is clear that conductivity, dielectric constant and other properties of composites shown at the left and right parts of Figure 1 are very different.

*msychov@yahoo.com.

The composite structure depends, among other factors, on interfacial interactions, namely on intensity of particle-particle, particle-polymer and polymer-polymer interactions. Such interactions depend on ability of active sites (or functional groups) of polymer and particle to attract corresponding active sites of particle or polymer, i.e., ability of being donor or acceptor. In chemistry such ability is also known as acid or base properties. Thus acid-basic properties reflect actually energy state of surfaces of interacting phases using common for chemists language. Below examples of such considerations are given for several composites.

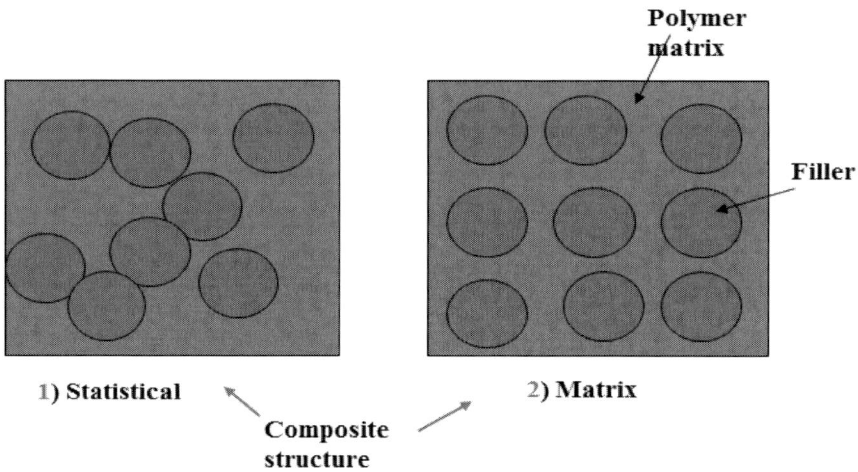

Figure 1. Nanocomposite structures.

Materials and Methods

Polymer composites were prepared using polymer solutions or latex. In the former case butadiene nitrile rubber (BNR), acrylic resin (AR) and poly(vinyl alcohol) cyanethyl ester (CEPVA) were used as binders. CEPVA structure is shown on the Figure 2.

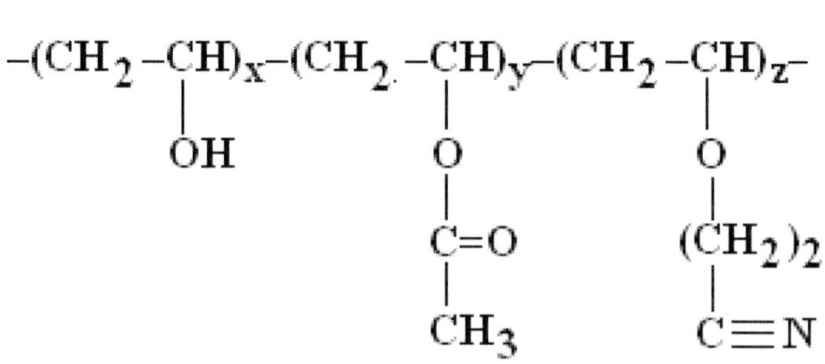

Figure 2. Structure of poly(vinyl alcohol) cyanethyl ester (CEPVA).

Polymers were dissolved in DMF (in case of CEPVA) or acetone (in case of BNR and AR) and then mixed with filler particles with desired ratio. In case of utilization of latex, powders were first ball milled in surfactant water solutions and then obtained suspensions were mixed with latexes. Butadiene nitrile (BN) and acrylate (BMNK and DMMA) latexes were utilized.

For dielectric composites several fillers were used: $BaTiO_3$ and two $BaTiO_3$ – $CaZrO_3$ ceramics (VS^{-1} and BCN). For conductive composites carbon blacks K354 and PM-80E as well as colloid graphite S-1 (specific surface area 1 m^2/g) were used as fillers. For measurements of dielectric constant composite films were Dr. Blade cast onto aluminum substrates and dried. Then Al electrodes were sputtered and electric properties measured using E7-20 RLC meter. For conductivity measurements films were cast on dielectric substrates with 4 electrodes and resistance was measured using B50 voltage source and E7-21 universal voltmeters.

Poly(vinyl alcohol) cyanethyl ester - $BaTiO_3$ Composites

In a first example dielectric composite is described comprising barium titanate ($BaTiO_3$) nanoparticles as a filler and poly(vinyl alcohol) cyanethyl ester as a polymer binder. In order to change composite's properties, surface acid-base characteristics of filler ($BaTiO_3$) were changed by various methods: heat-treatment, hydrolysis, utilization of surfactant, electron-beam treatment etc. [1].

For dielectric composites important characteristic is dielectric constant. Concentration dependences of dielectric constant of composites $\boldsymbol{\varepsilon}$ were fitted with Lichtenecker equation:

$$\varepsilon^k = \varphi_1 \varepsilon_1{}^k + \varphi_2 \varepsilon_2{}^k \quad (1)$$

where $\boldsymbol{\varepsilon_1}$, $\boldsymbol{\varepsilon_2}$ – dielectric constants of polymer and filer, φ_1, φ_2 – their volume fractions; $\boldsymbol{k}$ – constant. Figure 3 gives examples of fit, one can see that equation is quite suitable.

Acid-base properties of filler surface were studied by adsorption of indicator molecules [2]. Method is based on the measurements of amounts of specific surface centers (qpKa values, μmol/g) having certain values of the pK_a — dissociation constant characterizing acid–base nature of the center. The advantage of the method is that both differential (contents of specific acid–base centers) and integral (total amount of centers, acidity function H_o) characteristics of solid surface may be obtained [3].

In a number of experiments it was shown that dielectric constant of composite depends on amount of Broensted basic centers on surface of filler due to their interaction with acidic hydroxyls of polymer CEPVA [4-6]. For approximation of obtained results a modified Lichtenecker equation was introduced which gives good agreement between the experimental and calculated $\boldsymbol{\varepsilon}$ values:

$$\varepsilon^{a + b*IF} = \varphi_1 \varepsilon_1{}^{a + b*IF} + \varphi_2 \varepsilon_2{}^{a + b*IF} \quad (2)$$

where ***IF*** – parameter, reflecting intensity of interface interactions, in this case ***IF*** equals to amount of Broensted basic (BB) centers on surface of filler capable of interaction with acidic hydroxyls of polymer CEPVA. ***a*** and ***b*** are fit coefficients. Example of fit of experimental data with equation (2) is given in the Figure 4.

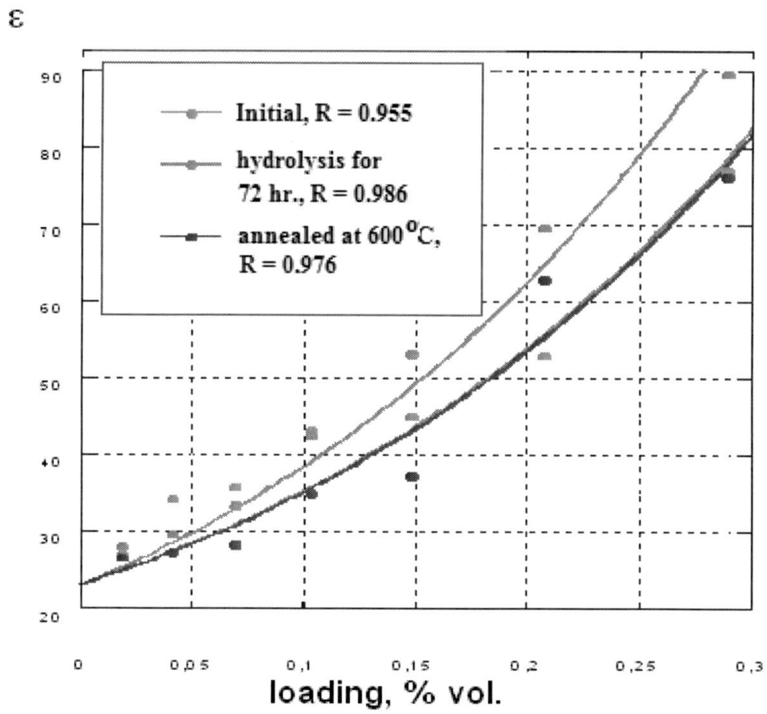

Figure 3. Dielectric constant of composite dependencies on concentration of fillers with different surface properties.

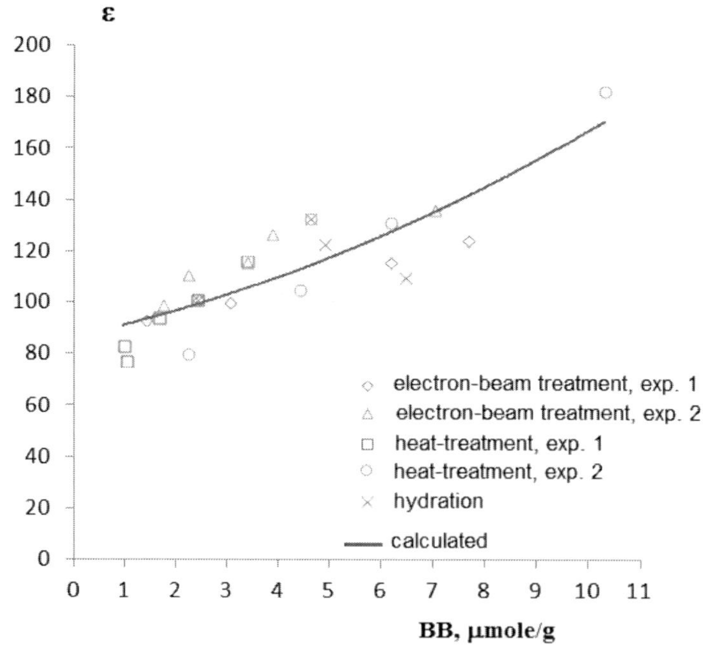

Figure 4. Example of fit of experimental data with equation (2).

Core-Shell Approach to CEPVA - $BaTiO_3$ Composites

Further experiments were continued utilizing core-shell approach. That means that barium titanate core was used to provide high dielectric constant and core was covered by several oxides in order to control surface acid-base properties of particles. TiO_2, Co_3O_4, Nb_2O_5, Co_3O_4+Nb_2O_5, WO, SiO_2, Al_2O_3, SiO_2+Al_2O_3, and Ta_2O_5 shells were formed by sol-gel method (treatment made in the lab of Prof. O.A. Shilova). MgO and ZrO_2 shells were done by plasma deposition (treatment made by Dr. A.A. Eruzin). Experiments confirmed validity of approach and equation (2). This equation is unique in having prognostic power to predict electric properties of composites depending on interfacial interactions.

Experiments with Other Binders

It should be noted that depending on polymer matrix, content of other active centers may be utilized in equation (2) depending on peculiarities of binder – solid interactions. Validity of approach was prooved using several ferroelectric fillers ($BaTiO_3$, VS-1, BCN) and binders (butadiene nitrile rubber BNR, acrylic resin AR), see Figure 5. When approximation curve reaches axis OY, i.e., surface contains no centers, value of k parameter in equation (1) reflects imaginary situation when surface of filler is absolutely inactive.

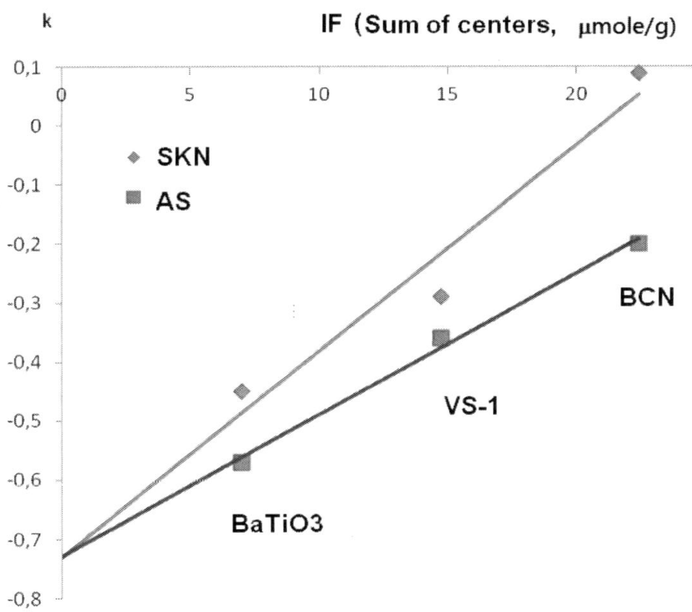

Figure 5. Dependence of value of k in equation (1) on surface properties of filler.

DIELECTRIC LATEX COMPOSITES

In the case of a latex (water dispersion of polymer) utilization of content of specific surface centers is not suitable due to complexity of the system – latex contains surfactants etc. In this case good results are obtained by introduction of new parameter

$$\Delta = pH_{latex} - H_o, \tag{3}$$

where pH_{latex} is pH of latex, H_o – acidity of filler's surface. In this case parameter Δ is used as a value of ***IF*** in the equation (2). Validity of approach was shown using several fillers (BaTiO3, ceramics VS-1) and latexes (butadiene nitrile BN, acrylate BMNK and DMMA), Figure 6.

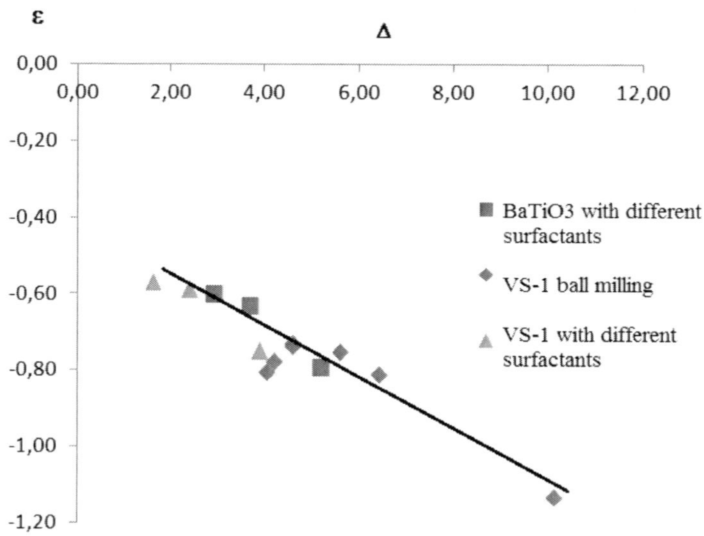

Figure 6. Interrelation between Δ and ε for latex composites.

CONDUCTIVE LATEX COMPOSITES

Further using butadiene nitrile latex BN and conductive carbon blacks K354 and PM-80E and graphite S-1 it was shown that the parameter Δ also is suitable in prognosis of electric conductivity of composites. Concentration dependencies of conductivity were approximated by equation of percolation theory [5]:

$$\sigma = \sigma_0 (c - c_n)^t , \tag{4}$$

where c – content of conductive filler, % vol.; c_n – percolation threshold, %vol.; t — index; σ_0 — constant. As it may be seen from Table 1, value of percolation threshold is in reverse correlation with value of parameter Δ. Consequently value of conductivity is in direct correlation with value of parameter Δ.

Table 1. Properties of conductive composites

Filler	Δ	c_n, % vol.	σ, S/m
K354	4,2	23	0,22
S-1	7,9	15,6	50
PM-80E	8,3	7,7	250

CONCLUSION

Thus, in a number of experiments it is shown that consideration of acid-basic properties of surface of filler and acid-basic interface interactions is a productive approach in explanation and prediction of composite electric properties. It may be expected that other properties sensible to such interactions in composite may be also predicted and promoted by control of acid-base properties of surface of filler, for example composite mechanical properties.

REFERENCES

[1] S. A. Alekseev, V. G. Korsakov, M. M. Sychov, O. V. Likhacheva, A.G. Rodionov, L.L. Ezhenkova, Effect of Donor–Acceptor Sites at a Barium Titanate Surface on the Properties of Composites Based on Cyanoethyl Polyvinyl Alcohol. *Rus. J. Appl. Chem.*, 2006, Vol. 80, No. 4, pp. 700–703.

[2] M. M. Sychov et al. Surface properties of ZnS and AC powder electroluminescent phosphors. *Journal of the Soc. for Information Display* 11 (1), p.33-38 (2003).

[3] M. M. Sychov et al. Study of active surface centers in electroluminescent ZnS: Cu,Cl phosphors. *Appl. Surf. Sc.* 244(1-4), 461-464 (2005).

[4] S. A. Alekseev, V. G. Korsakov, M. M. Sychov, O. V. Likhacheva, A. G. Rodionov, L. L. Ezhenkova, Effect of Donor–Acceptor Sites at a Barium Titanate Surface on the Properties of Composites Based on Cyanoethyl Polyvinyl Alcohol. *Rus. J. Appl. Chem.*, 2006, Vol. 80, No. 4, pp. 700–703.

[5] M. M. Sychov et al. // *Glass Physics and Chemistry*. 2011. V.37. No.5. P.734.

[6] M. M. Sychov et al. // *Glass Physics and Chemistry*. 2011. V.37. No.6. P.624.

[7] Aneli, J. N. *Structuring and conductivity of polymer composites* / J. N. Aneli, L. M. Khananasvili, G. E. Zaikov. – NY: Nova science publishers Inc., 2007. – P.37.

SHORT COMMUNICATIONS

Chapter 12

PROCEEDINGS FROM THE SEMINAR "NANOPHYSICS AND NANOMATERIALS"

This short communication section of the journal contains selected proceedings of the seminar "Nanophysics and Nanomaterials" hold in the University of Mines, St. Petersburg, Russia. This seminar with international participation contains reports from Europe, Asia and America in various branches of science related to nanotechnology.

Among university guests present at the seminar, there were the Research and Technology Director of Canadian Company CNT-MC Inc., the Board Director of North-West Technology Transfer Center, Vladimir Rudashevsky; the head of the special laboratory of Rosatom at the National Research Atomic University of MIPI Vadim Petrunin; main technologist of Ao "Benek" (Finland) Edgar Mayorov, the head of mining projects of IMC Montan (UK) Igor Pleskunov; Professor Evgeny Shadrin from Physical-Technical Institution (Ioffe) of RAS and many others.

Selected short communications will be included as full-sized papers in the next volume of the journal.

ABSORPTION SPECTRA AND OPTICAL PROPERTIES OF FLUOROPHOSPHATE GLASSES CONTAINING CUCL

M. Abdelghany, E. V. Kolobkova and M. M. Sychov*

St. Petersburg State Institute of Technology (Technical University)

Fluorophosphate glasses doped with noble metal and copper are of interest for solar power engineering, biosensors and light conversion in white LEDs. A series of fluorophosphates glasses of the composition [18 $NaPO_3$- 72 $Ba(PO_3)$- 10 AlF_3 mol %] doped with CuCl (1.72, 3.44, 5.16 and 8.6 mol% above 100 mol%) has been synthesized and characterized by UV-vis and near-IR spectroscopy. The spectra shown of glass samples with the average thickness 1.66 mm shown in Figure 1 indicate that the increase of CuCl content leads to the growth in the intensity of absorption peaks in the regions about 200-350 nm and 800-900 nm relating to Cu(I) clusters and Cu(II) ions correspondingly.

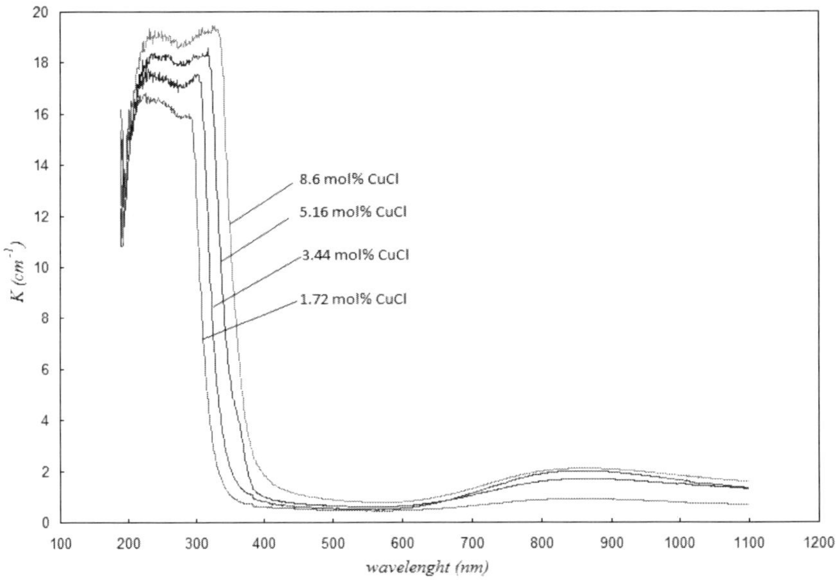

Figure 1. Absorption coefficient of 18 $NaPO_3$-72 $Ba(PO_3)_2$-10 AlF_3 mol% with different amount of CuCl.

On the basis of these spectra, the following characteristics of the studied glass samples were determined:

1. Cut-off wavelength (*λcut-off*) and optical band gap values (E_{opt}) according to Tauc's rule (Tauc, Mott and Davis power law) for the wavelength values corresponding to the

*E-mail: ph_abdelghany@yahoo.com.

absorption edge, where the intensity reaches the maximum value in optical absorption spectra are taken as cut-off wavelengths;

2. Refractive indices (n) according to Dimitrov-Sakka law (Dimitrov and Sakka, 1996; Eraiah, 2006) using the following equation:

$$\frac{n^2-1}{n^2+2}=1-\sqrt{\frac{E_{opt}}{20}}$$

As shown in Figures 2 and 3, cut-off wavelength and refractive index values grow while the optical band gap drops with CuCl content due to structural changes in the glass and formed Cu-based clusters.

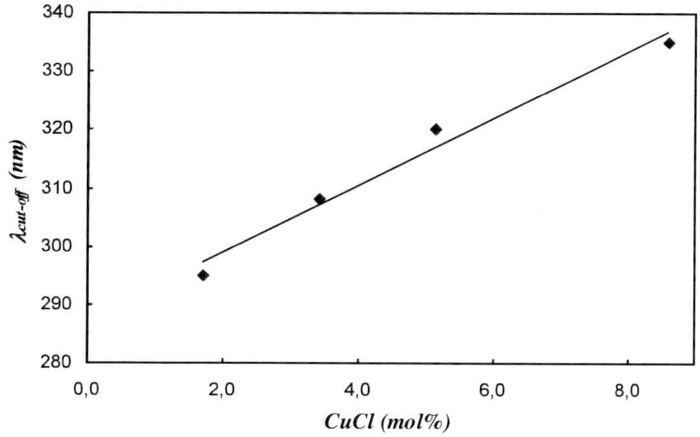

Figure 2. Cut-off wavelength of the studied glasses as a function of CuCl content

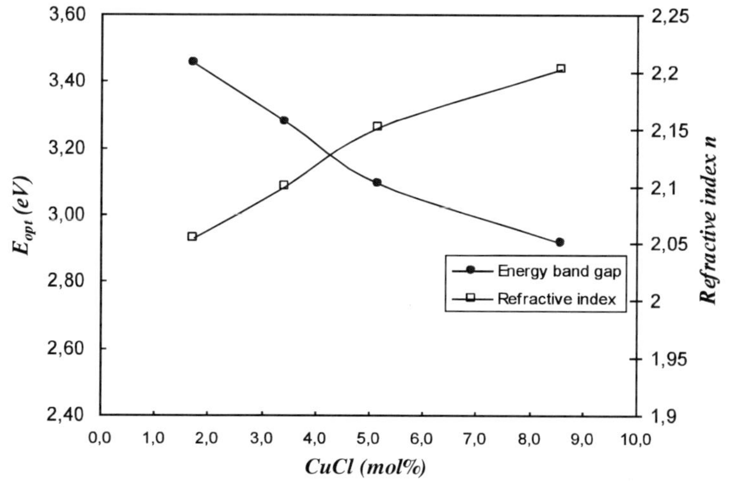

Figure 3. Energy band gap values and refractive index of the studied glasses as a function of CuCl content.

CONCLUSION

Generally, the considered approach provides a possibility to obtain glasses with adjustable optical characteristics due to small changes in their composition.

REFERENCES

Davis.E.A. and Mott. N.F , 1970, Conduction in non-crystalline systems V. Conductivity, optical absorption and photoconductivity in amorphous semiconductors, *Philosophical Magazine,* Vol., No.179, pp. 903-922.

Dimitrov.V and Sakka.S, 1996, Electronic oxide polarizability and optical basicity of simple oxides-I, *J. Appl. Phy*, Vol.79, No.3, pp.1736-40.

Eraiah.B, 2006, Optical properties of samarium doped zinc–tellurite glasses, *Bull. Mater. Sci.,* Vol. 29, No. 4, pp. 375–378.

Mott, N.F. and Davis, E.A, 1971 "*Electronic Processes in Non-Crystalline Solids*", Oxford University Press, London.

Murugavel.S and Roling.B, 2007, Ion transport mechanism in borate glasses: Influence of network structure on non-Arrhenius conductivity, *Physical Review* B, Vol. 76 pp. 180202-5.

S.Lakshimi Srinivasa Rao, G. Ramadevudu, Md. Shareefuddin, Abdul Hameed, M.Narasimha Chary and 3M.Lakshmipathi Rao, *International Journal of Engineering, Science and Technology*, Vol. 4, No. 4, 2012, pp. 25-35.

USE OF THE MIEDEMA'S MODEL TO PREDICT THE STANDARD ENTHALPY OF FORMATION OF THE INTERMETALLIC CADMIUM-TIN NANOPARTICLE

Nicolay Barbin[*,1,2], *Sergey Alexeev*[2,3], *and Irina Ovchinnikova*[4]
[1]Chemistry Department, Ural State Agrarian University, Yekaterinburg, Russia
[2]Scientific Department, Ural Institute of State Fire Service of Emercom of Russia, Yekaterinburg, Russia
[3]Science and Engineering Centre "Reliability and Safety of Large Systems" of Ural Branch of Russian Academy of Sciences, Yekaterinburg, Russia
[4]Physics Department, Ural Institute of State Fire Service of Emercom of Russia, Yekaterinburg, Russia

ABSTRACT

The Miedema's model is used for calculation of the standard enthalpy of formation of the intermetallic cadmium-tin nanoparticle. $\Delta_f H^0_{289}$ is 1.29 kJ/mole.

Keywords: Miedema's model, calculation, standard enthalpy of formation

The structure of the Cd–Sn alloy contains intermetallic nanoparticles. One intermetallic compound has formed in this alloy [1]. The standard enthalpy of formation is one of essential thermodynamic characteristics. Its experimental definition is laborious and complicated problem therefore calculation methods are successful alternative of this problem solution.

In the Miedema's model [2-4], the enthalpy of formation of Cd–Sn alloy $\Delta_f H^0_{289}$ is represented by

$$\Delta_f H^0_{289} = f\left(C^S_{Cd}, C^S_{Sn}\right) g\left(C_{Cd} C_{Sn}\right) FP - \left[\left(\Delta\varphi^*\right)^2 + \frac{Q_0}{P}\left(\Delta n^{\frac{1}{3}}_{WS}\right)^2 - \frac{R}{P}\right] \quad (1)$$

where $f\left(C^S_{Cd}, C^S_{Sn}\right)$ and $g\left(C_{Cd} C_{Sn}\right)$ are functions of the component concentration. P, $\frac{Q_0}{P}$, $\frac{R}{P}$ are empirical constants for various groups of alloys. φ^* is the index of electronegativity.

*Corresponding author: Prof. Nicolay Barbin. Head of Chemistry Department, Ural State Agrarian University, Yekaterinburg, Russia. Tel.: + 7343 2214103. E-mail: NMBarbin@mail.ru.

Positive number $\frac{Q_0}{P}\left(\Delta n_{WS}^{\frac{1}{3}}\right)^2$ is considered influence of difference of electron density at the boundary of the Wigner-Seeitz atomicells of pure metals. V is the molar volume of an atom in a complex compound.

Parameters for the calculation of the Miedema's model are listed in Table 1.

Table 1. Parameters for the calculation of the Miedema's model

Parameter	φ^*, V	$\Delta n_{WS}^{\frac{1}{3}}$, (Unit area)$^{1/3}$	V$^{2/3}$, cm^2	Q/P, V^2/(Unit area)$^{2/3}$
Sn	4.15	1.24	6.40	1.00
Cd	4.05	1.24	5.50	1.00

Surface concentrations of Cs and Sn in CdSn are counted by formulas (2) and (3) [2-4].

$$C_{Cd}^s = C_{Cd} V_{Cd}^{\frac{2}{3}} / \left(C_{Cd} V_{Cd}^{\frac{2}{3}} + C_{Sn} V_{Sn}^{\frac{2}{3}} \right) \tag{2}$$

$$C_{Sn}^s = C_{Sn} V_{Sn}^{\frac{2}{3}} / \left(C_{Cd} V_{Cd}^{\frac{2}{3}} + C_{Sn} V_{Sn}^{\frac{2}{3}} \right) \tag{3}$$

Table 2 summarizes the calculated values of atomic and surface concentrations, and molar volumes of atoms.

Table 2. Results of calculation of concentrations and molar volumes of Cd and Sn in CdSn

Nanoparticle	C_{Cd}	C_{Sn}	C_{Cd}^s	C_{Sn}^s	$V_{Cd}^{\frac{2}{3}}$, cm^2	$V_{Sn}^{\frac{2}{3}}$, cm^2
CdSn	0.50	0.50	0.46	0.54	5.55	6.4

where C_{Cd} and C_{Sn} are atomic concentrations of Cd and Sn in CdSn. C_{Cd}^s and C_{Sn}^s are surface concentrations of Cs and Sn in CdSn. V_{Cd} and V_{Sn} are molar volumes of Cd and Sn in CdSn.

Functions of the component concentration are obtained by formulas (4) and (5) [2-4].

$$f\left(C_{Cd}^S, C_{Sn}^S\right) = C_{Cd}^S \times C_{Sn}^S (1 + 8\left(C_{Cd}^S \times C_{Sn}^S\right)^2) \tag{4}$$

$$g\left(C_{Cd} C_{Sn}\right) = 2\left(C_{Cd} \times V_{Cd}^{\frac{2}{3}} + C_{Sn} \times V_{Sn}^{\frac{2}{3}} \right) / \left(V_{Cd}^{\frac{2}{3}} + V_{Sn}^{\frac{2}{3}} \right) \tag{5}$$

The matching way [5] are used for calculation of P, Q/P and R/P. $\left(\Delta\varphi^{*}\right)^{2}$ and $\left(\Delta n_{WS}^{\frac{1}{3}}\right)^{2}$ are found by the additive method and using data of table 1.

Table 3. Calculated values for P, Q/P, R/P, $\left(\Delta\varphi^{*}\right)^{2}$ and $\left(\Delta n_{WS}^{\frac{1}{3}}\right)^{2}$

Nanoparticle	$\left(\Delta\varphi^{*}\right)^{2}$, V	$\left(\Delta n_{WS}^{\frac{1}{3}}\right)^{2}$	R/P, V^2	P, $V^{-1}\times cm^{-1}$	f	G
CdSn	0.01	0	0.40	0.25	0,37	1.00

Prediction by the formula (1) is given 1.29 kJ/mole for the standard enthalpy of formation of the intermetallic cadmium-tin nanoparticle.

REFERENCES

[1] LeBlanc M., Naumunn M., Tschesno D., *Math. Phys.,* 79, (1927) 72-106.
[2] Miedema A.R., J. *Less Common. Met.,* 32(1), (1973) 117-136.
[3] Miedema A.R., Boom R., De Boer F.R., J. *Less Common. Met.*, 41(2), (1975) 283-298.
[4] Miedema A.R., J. *Less Common. Met.*, 46(1), (1976) 67-83.
[5] Ovchinnikova I.V., Terentev D.I., Alexeev S.G., Barbin N.M. *Rasplavy* (Melts), (5), (2011) 83-91 (russ.).

APPLICATION OF ELECTROCHEMICAL IMPEDANCE SPECTROSCOPY FOR CHARACTERIZATION OF NANOPOROUS FILMS

***A. V. Bezmaternykh*[1], *K. L. Levine**[*,1], *A. G. Syrkov*[1], *I. V. Pleskunov*[2] *and V. V. Afanas'ev*[3]**

[1]University of Mines, St. Petersburg, Russia
[2]IMC Montan, London, Great Britain
[3]University of Leuven, Leuven, Belgium

ABSTRACT

Electrochemical impedance spectroscopy is a non-destructive method for films characterization. Spectra of electrochemical impedance in encrypted way carry valuable information about thin film physical properties. Revealing this data is a non-routine problem which requires alliance of modeling, computing and careful experimental advances. This short communication provides a brief overview of EIS method with an emphasis of solving mentioned practical problems of thin film characterization.

Keywords: diffusion, electrochemical impedance spectroscopy, porous materials, Nyquist plot, Bode plot, electrochemical impedance

INTRODUCTION

In EIS, variable frequency potential is applied to an electrode covered with the studied sample. Impedance measurements are taken in the solution of an electrolyte. Collected data is presented in two types of graphs: Nyquist and Bode plots which are the imaginary part of impedance vs. real part of impedance, and impedance modulus vs. frequency logarithm, respectively.

Description of the method

EIS allows studying double electric layer at electrode/solution interface when no direct measurements are possible. Nyquist plot is shown in Fig 1. In the particular case shown in this figure, the plot consists of a part of semicircle, which is indicative to double electric layer or interphase, and part of a straight line. Equivalent circuits constructed by fitting EIS data contains passive circuit elements: resistor, capacitors, constant phase element and Warburg element. In the equivalent scheme shown in Fig. 2, resistor accounts for solution resistance; the capacitor for double electric layer capacitance, therefore indicating the interface; Warburg

[*]Corresponding author: levinkl@hotmail.com.

element (W.) shown in this figure is an empirical element, which does not have a simple analogue in electronics, It is responsible for diffusion.

W. element allows obtaining main characteristics of the diffusion. If real part of the impedance is plotted at abscises axes, and imaginary part is plotted at ordinate axes, line at 45 degrees at the right hand side of a semicircle is indicative for the diffusion. This line can be fitted by W [1] and establishes correlation between parameters of uniform diffusion and pore size. Solving practical problems, however, non-uniform diffusion frequently takes place. This is the diffusion through ensemble of pores of variable size conditionally varying by length and size. Theoretical basis for solving the mentioned problem suggests [2] with the equation:

$$Z\_0 = [\![(ZR)]\!]^{\wedge}(1/2)\ \mathrm{cotan}h[(l]\ ,$$

where Z_0 measuring impedance, Z impedance of film/solution interface, R ohmic resistance of the interphase, l pore depth, ρ quantity reciprocal to so-called penetration length, which can be linked to diffusion coefficient.

Systematically varying characteristics of pores, electrolyte resistance in pores obtained by electrochemical methods (pore resistance) [3]) can be linked with the characteristics of the diffusion through studied structure and allows to establish correlation between measured characteristics of the diffusion with microscopic characteristic of a sample.

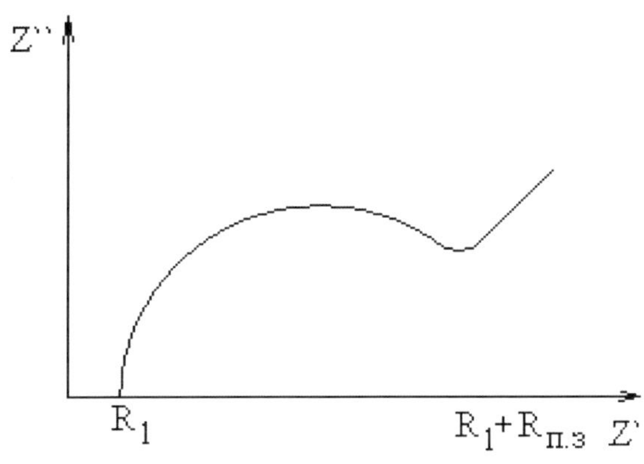

Figure 1. Typical Nyquist plot.

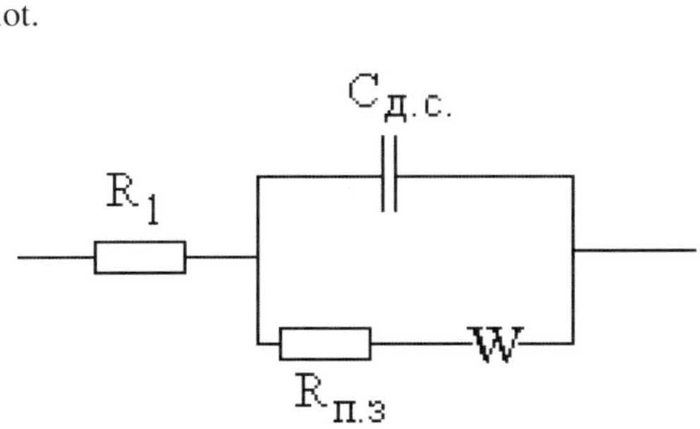

Figure 2. Electrical equivalent schematic fitting of Nyquist plot shown in Figure 1.

CONCLUSION

EIS data contains valuable information about characteristics of nanoporous films. Applying relevant experiment al methods and data analysis allows extracting data about the diffusion through nanoporous film, pore size distribution, diffusion through film different types of ions.

REFERENCES

[1] E.G. Tolstopyatova, S.N. Sazonova, V.V. Malev1, V.V. Kondratiev, Electrochemical impedance spectroscopy of poly(3-methylthiophene) and poly(3-octylthiophene) film electrodes, *ElectrochimicaActa,* 50, (2005), 1565-1571.

[2] Robert De Levie, Electrochemical response of porous and rough electrodes, *Advances in electrochemistry and electrochemical engineering*, Vol. 6, 1961, pp 329-397.

[3] K.L. Levine, J.O.Iroh, Resistance of the Polypyrrole/Polyimide Composite by Electrochemical Impedance Spectroscopy, *Journal of Porous Materials* 11: 87–95, 2004.

IODIDE TRANSPORT - METHOD OF SYNTHESIS OF INORGANIC MATERIALS

S. P. Bogdanov[*]
St. Petersburg Technological Institute
(Technical University)

Keywords: iodide transport, inorganic materials, coatings

INTRODUCTION

Most of the reactions of the «solid–solid» type are «solid–gas–solid» reactions, where the gas phase is present as a transporting agent. Many examples of such reactions are known [1]. The characteristic feature of the gas-transport reactions is the interaction of the solid (liquid) substance with any gaseous substance with the formation of a gaseous product that is decomposed with the precipitation of the original substance after being transported to another part of the system.

Among the known substances, comfortable and affordable agent, satisfying to this condition, is iodine. However iodine-transport now in industry id used only for refining of metals [2].

During last five years, our group was successfully studying gas-transport reactions with iodine participation.

RESULTS AND DISCUSSION

It is revealed, that the iodine-transport method can be used for synthesis of a wide range of inorganic materials [3-7].

1. Producing Coatings on Powder Materials

In cited papers, reactor without a special gradient of temperatures was used. In this case, driving force of the reaction is the shift of chemical balance to the side of the formation of the compound on the solid surface. This method allows receiving coating on powders with the size of particles from ten nanometers up to several millimeters. The thickness of coatings makes from 1-2 nm up to several micrometers.

The powders of diamond, boron nitride, tungsten, carbide, covered by metals: titanium, chromium, cobalt, nickel, iron and their compounds was obtained for manufacturing tools.

[*]Corresponding author: Assoc. Prof. Sergey P. Bogdanov. St. Petersburg Technological Institute (Technical University). Moskovskii pr. 26, St. Petersburg, 198013 Russia. E-mail: BogdanovSP@mail.ru.

The coatings activate sintering of a a composite, allowing in regular intervals to form uniform distribution of binders. The special coatings can stop recrystallization of a grain during sintering.

Layer of metals and alloys monocrystals of diamond and silicon carbide is offered, they are used for manufacturing of the ceramics for heat conduction.

Thin films on a surface of powders have high chemical activity. Layers of metals oxides with a high specific surface can be used as catalysts.

2. Diffusion Saturation of a Surface of Metals and Alloys

The iodine-transport method allows not only transferring metal or some metals on a surface of a substrate, but also implanting them in a material. The use of iodine in this case a variant of a known method of diffusion saturation metals [8]. This process allows to functionalize surface by increasing its chemical resistance, hardness, wear resistance, etc.

Implantation of alloying elements with the help iodine-transport opens an opportunity of synthesis of various types of semi-conductor materials, including luminophores.

3. Synthesis of Powders of Inorganic Materials

The iodine opportunities both as transport agent and as catalyst of chemical reactions are realized in a method. The metals transferred iodide-transport; at the moment of allocation (or iodides) have the greater activity, than initial metal powders. They can enter chemical reaction with a material of a substrate at lower temperatures, than what is necessary in iodine absence.

The offered method of powder synthesis of solid solutions of metals and intermetallides allows to make synthesis at temperatures below than melting temperature. These materials are interesting to creation cermets and coatings. In the work some bi-metal systems are investigated: Al-Co, Al-Cr, Al-Cu, Al-Ni, Al-Ti, B-W, Co-Ti, Cr-Cu, Cu-Ti, Cu-Si, Ni-Ti, Si-Ti, Si-W.

Iodide-transport activates interaction of metals with carbon, silicon, boron, nitrogen, that has allowed to synthesize variety of carbides, silicides, borides and nitrides at temperatures much lower than usually used for industrial obtaining of those materials.

To synthesize titanium carbide TiC from metal titanium and soot in the presence of iodine, there is enough 400^0C, comparing to industrial manufacturing utilizing heating up to 1600^0C [9]. TiN in powder form in the presence of iodine is formed already at 250^0C instead of 1200^0C at usual process [10]. Above 450^0C titanium iodide converts to gas phase. The reaction between two gases: TiI_4 and N_2 results in the formation of TiN nanoparticles. As a result, nitriding titanium particles by the size up to 10 microns are present in a sample, as well as TiN particles by the size 20-200 nm are present in a gas-phase.

4. Speeding-Up Exchange Reaction

The method can be used for acceleration of chemical reactions of an exchange with participation of metals. An example of such processes is reduction of oxides by aluminothermy. It is established that for a full recovery of SiO_2 there 600^0C is enough. By overlapping in one reactor two processes of iodide-transport - recovery and refining can be solved a task of manufacturing of silicon of high purity.

Simplicity of this method, requiring inexpensive equipment makes it a promising technological process.

REFERENCES

[1] Schafer H. *Chemical transport reactions*, Germany: Wiley, 1962.
[2] Rolsten R.F. *Iodide Metals and Metal Iodides*, New York: Wiley, 1961.
[3] Bogdanov S.P. *Glass Phys. Chem*., 37, (2), (2011) 172–178.
[4] Bogdanov S.P. *Glass Phys. Chem*.,.38, (6), (2012) 750-754.
[5] Bogdanov S.P. *Bulletin of the Saint Petersburg State Institute of Technology* (Technical University), 16(42), (2012) 24-28.
[6] Bogdanov S.P. *Glass Phys. Chem.,* 39, (4), (2013) 444–447.
[7] Bogdanov S.P. *Glass Phys. Chem*., 39, (4), (2013) 448–452.
[8] Radomiselsky I.D., Napara-Volgina S.G. *The getting alloyed powders diffusion method and their usage*, Kiev, Naukova Dumka, 1988.
[9] Kiparisov S.S., Levinsky Y.V., Petrov A.P. *Titanium carbide (production, properties,* application), Moscow, Metallurgy, 1987.
[10] Samsonov G.V. Nitrides, Kiev, *Naukova Dumka*, 1969.

By Applying the Method Equal Channel Angular Extrusion for Products Submicron Structure at the Bulk of Metal

V. I. Bolobov and S. A. Chupin*

Department of Mechanical Engineering,
National Mineral Resources University, Russia

Abstract

Influence of grain size on the mechanical and performance properties of metallic materials is analyzed. Methods for grain refining- equal channel angular extrusion (ECAE) is reviewed in the article. The method of equal channel angular pressing is described. Experimental data on the effect of sub-micron structures on hardness after ECAE samples of copper, aluminum and steel used in the manufacture of mining cutters (35HGSA) is presented.

Keywords: sub-micron structure, grain metal, mechanical properties, equal channel angular extrusion, mining cutters

Introduction

Grain refinement is one of the most effective ways to improve the full range of mechanical and performance properties of metallic materials: hardness, strength, toughness, wear resistance, etc. For example, dependence of increase yield strength and grain size of the metal is expressed in the equation of the Hall-Petch

$$\sigma_{0,2} = \sigma_0 + k \cdot \frac{1}{d^2}$$

σ_0 - shear stress required for dislocation slip in a single crystal, MPa; k - ratio of Hall-Petch; d - grain size, m.

Existing methods of grinding (modification melts, recrystallization, grain recovery, metal forming) are limited to the achievement of the grain size of several microns. Nowadays there are new methods of grinding grain, allowing the material to give the nano and submicrocrystalline structure such as powder technology, severe plastic deformation, film technique [1]. Severe plastic deformation copper samples with initial grain size of 50 microns results in grain refinement to 0.28 microns with an increase of tensile strength and yield strength of the material and 4.2 to 3 times, respectively [2]. The same effects on samples of

*Corresponding author: Prof. V.I.Bolobov. Doctor of Sc. Department of Mechanical Engineering, National Mineral Resources University, Russia. E-mail: Boloboff@mail.ru.

aluminum alloy refine grain with 20 up to 0.2 micron, which increases the tensile strength of 1.3 and a yield of 1.6 times [3]. At intensive plastic deformation of pipe steel 20G2SF grain is milled from 8 to 0.63 microns, with a corresponding increase in hardness and yield strength at 1.2, 2 and 1.6 times [4].

For bulk products with small grain size is using equal-channel angular pressing (ECAP) as the most appropriate method of severe plastic deformation [5]. When implementing this method, the metal blank is repeatedly pressed in a special matrix via two channels with the same cross sections which intersect at the angle of 90 - 135 °, at room or elevated temperatures depending on the deformability of the material. After each pass, the preform is rotated around its longitudinal axis by an angle of 180 °.In the first pass, the cell of metal takes the form of an ellipsoid because of the shift in the intersection of the channels. Repeated passage leads to a shift in the same plane but in the opposite direction, causing cell re-acquires a spherical shape [6]. Electron microscopic study of the dislocation structure of metals after ECAE shows that these large strains cause fragmentation of grains of material with the appearance of these large-angle disoriented blocks of submicron size d = 100-400 nm [7-10].

A variation of this method is a method of holding it in a high temperature and the use of ECAP operation after hardening material.

Materials and Methods

Mining University is working to improve mechanical and performance characteristics of shell rock cutting tools by giving sub- and nanocrystalline structure to their body. Special device was developed to obtain materials with this structure (Figure 1a), which consists of: - two massive steel plates with intersecting at an angle of cylindrical channels in 135° (Figure 1b), punch and tightening bolts and nuts. The diameter of the punching samples was 8 mm and length 30 mm.

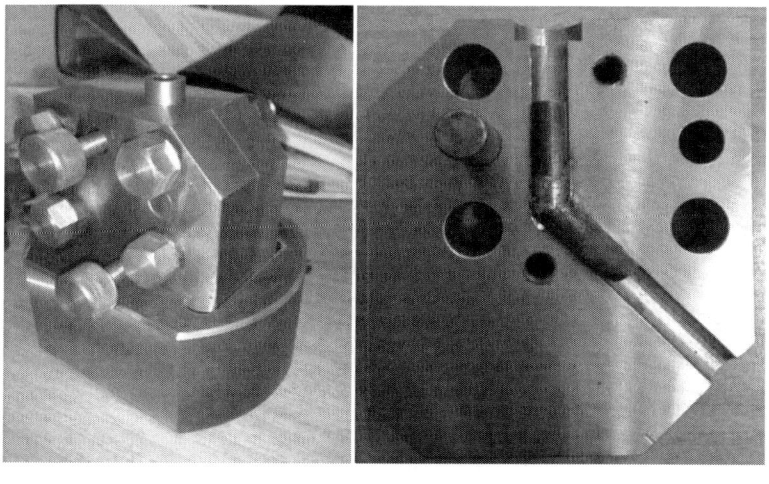

Figure 1. Devices for giving bulk samples of metals sub-nano-crystalline structure (*a*) and the inner surface of the sample matrix with the punching steel (*b*) is illustrated.

Tinius Olsen press force of 100 tons was used for punching sample.

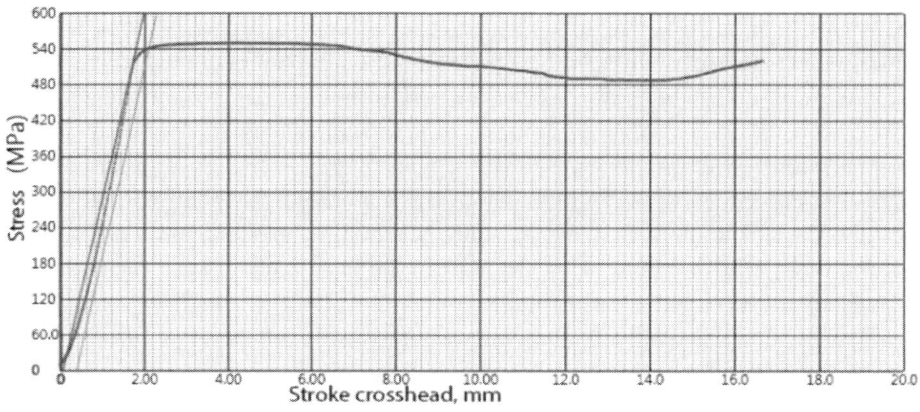

Figure 2. Dependence of the stress on the stroke crosshead press at punching copper sample through the channel.

Figure 2 shows that the stress at which the sample is forced through the channel is almost constant in time. This may indicate that the vast majority of stress of punching is fall to the bending of sample at the intersection of channels and there is only a small part is spent to due to the friction of the metal sample of the channel wall.

The object of investigation were chosen technical copper and aluminum as easily deformable materials, as well as 35HGSA steel which commonly used for the manufacture of the cutter body.

RESULTS AND DISCUSSION

In accordance with the results of the first experiment it was foundthat the grain refinement from 10 microns to 0.5 microns (the expected size) by equal-channel angular extrusion leads to a significant increase in hardness of analytes: Steel 17% (after 1 passage), copper 30% (after 2 passes), aluminum of 44% (after 2 passes). In accordance with existing dependencies binding hardness metals with other physico-mechanical characteristics, such 35KhGSA increasing steel hardness corresponds to an increase of its strength properties in the resulting ECAP 20%. By analogy with that of other metals, it can be expected that increasing the number of passes in the ECAP allow to obtain the nanocrystalline structure of metallic material being analyzed with a corresponding increase of their mechanical and performance properties.

In terms of research - the study of the structure and grain size, as well as wear resistance and strength properties of metals and steel analyte resulting from equal-channel angular extrusion at a higher number of passes.

Conclusion

- The influence of grain size on the physical and mechanical properties of metallic materials is analyzed;
- A design of matrix which allows to press the samples of metallic materials through crossing channels is developed;
- Samples of 35HGSA steel, copper and aluminum with improved hardness are obtained as a result of ECAP;

References

[1] Andrievskiy R.A. Nanomaterials: Concept and Current Issues / / *Russian Journal of Chemistry* (Journal of the Russian Chemical Society. Mendeleev), 46(5), (2002) 50-56.

[2] G.I. Raab, N.A. Krasilnikov, R.Z. Valiev, Processing Ultrafine-Grained Copper by ECAP with Controlled Back-Pressure, Ultrafine Grained Materials III, *Proceedings of Symposium held during 2004 TMS Annual Meeting in Charlotte*, USA, March 14-18, 2004, p. 137-143.

[3] Valiev, R.Z., Krasilnikov, N.A., Tsenev, N.K. Plastic Deformation of Alloys with Submicro-Grained Structure. *Mater. Sci. Eng.* A137, 35-40 (1991).

[4] Dobatkin S.V., Odesskiy P.D., Pippan R. Warm and hot equal channel angular extrusion of low-carbon steels // *Metals,* 1, (2004) 110-119.

[5] Valiev R.Z. *Nanostructured materials produced by severe plastic deformation* / Valiev R.Z., Alexandrov I.V. Logos, M. (2000) 272.

[6] Segal V.M., Reznikov V.I., Kopylov V.I. Plastic processes structuring metals. - Minsk: *Navukaitehnika,* (1994). 250.

[7] J. Gubicza, N.Q. Chinh, Z. Horita, T.G. Langdon. *Mater. Sci.Eng.* A 387/389, 55 (2004).

[8] Q. Liu, X. Huang, D.J. Lloyd, N. Hansen. *Acta Mater.* 50, 15,3789 (2002).

[9] A. Goloborodko, O. Sitdikov, R. Kaibyshev et al. *Mater. Sci.Eng.* A 381, 121 (2004).

[10] A. Belyakov, T. Saki, H. Miura, K. Tsuzaki. *Phil. Mag.* A 81,11, 2629 (2001).

SYNTHESIS AND QUANTUM CHEMICAL ANALYSIS OF PHOSPHORUS-TITANIUM OXIDE NANOSTRUCTURES ON SILICA SURFACE

E. O. Drozdov and S. D. Dubrovensky
St. Petersburg State Institute of Technology,
St. Petersburg, Russia

Silica supported multicomponent nanoscale systems, including phosphorous-titanium oxide ones, are shown to have promising applications in areas of catalysis and sorption. One of the most effective techniques for producing such systems is molecular layering (ML) method that is carried out on a surface of a solid support [1]. Herewith, a significant number of local chemical reactions takes place. In this connection, effective ways of both identification of products and determining influence of synthesis conditions on the composition and structure of formed coatings have to be found. Low dimensional nature of such systems gives us an opportunity to use quantum chemical methods for this objective.

The main purpose of this study was building and comparative quantum chemical analysis of phosphorus and titanium containing system models, including two-component phosphorus-titanium oxide one, selection of optimal conditions and executing the synthesis of mentioned silica supported monolayer nanocoatings.

Quantum chemical calculations were carried out by using the Gaussian®09 quantum chemical package at B3LYP/6-31G(d,p) level of theory that allows prediction structural characteristics and harmonic frequencies of considered structures with sufficient accuracy [2]. Cluster models of phosphorus containing structures with the number of Si-O-P bonds (denticity, functionality) varying from 1 to 3 were calculated.

$$n(\equiv Si-OH)+POCl_3 \rightarrow (\equiv Si-O)_n POCl_{3-n} + nHCl, \quad (1)$$

$n = 1\text{-}3$ (functionality)

Hydroxyl species of silica surface were represented by cluster model of minimal size (H_3SiOH) as it was earlier shown that it can be used to successfully predict structural and vibrational characteristics of surface groups [2]. Hydrogen atoms were used as pseudoatoms. An estimation of structural strains that originate during formation of polyfunctional species varies from 30 to 90 kJ/mol as follows by the comparing cluster models including from 1 to 10 silicon atoms.

Thermodynamic prognosis of equilibrium composition of one-component coating formed during chemisorption phase was carried out on the basis of quantum chemical calculations data. Stepwise increase of surface species denticity with the growth of the temperature is expected for both modificators. At low temperature, the formation of monofunctional titanium containing surface groups is relatively higher than one of monodentate phosphorus containing structures. At the same time, polyfunctional phosphorus oxide species are more stable than titanium containing ones at high temperature.

Phosphorus and titanium containing coatings on silica surface were synthesized using ML method at 200 °C in dry air flow of $POCl_3$ and $TiCl_4$ vapors. FTIR spectra of samples were further obtained and chemical analysis of [P], [Ti] and [Cl] content was carried out. The absorption band of phosphorus containing silica observed at 1012 cm^{-1} is referred to the stretching Si-OP mode. This value fully corresponds with quantum chemical calculations. According to the chemical analysis data, both bi- and tridentate phosphorus oxide groups were formed.

For titanium oxide coating, the absorption band at 960 cm^{-1} was observed in FTIR spectrum. According to quantum chemical predictions it can be referred to the stretching Si-OTi mode in polydentate species. Chemical analysis data let us to conclude that mostly bidentate titanium containing groups were present in the sample. Thus the opportunity of determining the presence of polyfunctional species in titanium oxide coatings using IR spectrum was shown. Results obtained for one-component coatings were used further for comparison with phosphorus-titanium containing system.

Such materials were synthesized by sequential treatment of silica surface with reactants vapor [1]. In that case, the ratio of components is defined by time. We propose another approach which supposes an interaction between surface active sites and a mixture of reactant vapors. So the composition of coating is controlled by the ratio of concentrations of gaseous reagents. Experimental search for demanded ratio of components for such systems is a very time-consuming process. Thus the usage of quantum chemical modeling results can be considered as a perspective method for this purpose.

The process may be complicated by donor-acceptor interactions between phosphorus- and titanium containing sites. Quantum chemical calculations show that such interaction is of high thermodynamic probability at low temperatures when monofunctional titanium oxide species take part in it.

Thermodynamic prognosis of equilibrium composition of the two-component coating depending on synthesys conditions was carried out. If steric barriers are absent, the main amount of titanium would be expected to be part of donor-acceptor complexes of monofunctional titanium oxide structures and adjacent polyfunctional phosphorus containing ones. While dehydroxylation the distance between neighbor surface species increases and leads to formation of isolated monodentate groups. So it is expected that Ti/P ratio would increase under these conditions along with temperature rise.

Samples of phosphorus-titanium containing coating on the silica surface were synthesized by chemisorption of reactants vapor mixture. Silica was pretreated at 200 °C (P-Ti-200) and 600 °C (P-Ti-600). It is shown on the basis of chemical analysis and IR spectroscopy that titanium is located on the surface in the form of mostly monodentate species in both series of samples. At the same time, the denticity of phosphorus oxide groups varies from around 2 in P-Ti-200 series to around 1 in P-Ti-600 ones. Phosphorus to titanium ratio qualitatively corresponds to the results of quantum chemical calculations. Presence of donor-acceptor interactions in solid phase was detected by comparing the electronic diffuse reflectance spectra of one-component and two-component coatings.

The results obtained confirm the probability of using quantum-chemical modeling for both selection of conditions for synthesizing multicomponent coatings by ML method and interpreting their spectral characteristics. Thus an effective solution to the problem of regulating the composition of nanoscale coatings is provided.

REFERENCES

[1] Malygin, A.A. Russ. J. *Gen. Chem.* 2002, vol. 72, № 4, p. 617 [in Russian].
[2] Malygin, A.A., Dubrovensky, S.D. *Russ. Chem. J.* 2009, vol. LIII, № 2, p. 98-110 [In Russian].

NANOTECHNOLOGY OF LOW-DIMENSIONAL OXIDE AND NITRIDE SYSTEMS

Yu. K. Ezhovskii
St. Petersburg State Technological University,
St. Petersburg, Russia

Progress in chemical nanotechnology using matrix synthesis of low-dimensional systems largely depends on the choice of optimum reagents and conditions for surface reactions. In molecular lamination (ML) [1, 2] or atomic layer deposition (ALD) [3–5] processes, the surface functional groups react with the low-molecular reagent, forming surface compounds that are the structural units of the material (e.g., a thin film) being synthesized.

The process is self-regulating and allows the composition and thickness to be controlled to an accuracy of a monolayer.

In the quantitative approach based on the inductive effect in the surface chemistry of solids [6], the reactivity of the functional groups of the surface is evaluated from the inductive parametric constants of the solid framework. This makes possible to determine the route and stoichiometry of surface reactions and adjust the temperature and composition of the reaction medium, the contact time of reagents, etc.

An analysis of the experimental data on the reactivity of the hydroxyl groups of the silica, silicon, and gallium arsenide matrices suggests that on the real surface, the hydroxyl groups are characterized by an activity that depends on the background of the sample, namely, the thickness of the oxide layer formed after preliminary processing. An analysis of the induction effect revealed a quantitative dependence of the reactivity of the hydroxyl groups on the surface of silicon ($\sigma^*_{S(Si)}$) on the thickness of the oxide layer (d_c) [7, 8]. An increase in the parametric constant from $\sigma^*_{S(Si)}$= 2,9 for the freshly etched silicon surface containing an approximately 1 nm oxide layer to $\sigma^*_{S(Si)}$= 4.57 characteristic of silicon dioxide at d_c>3 nm indicated that the activity of hydroxyl groups increased with the thickness of the oxide layer.

The comparatively weak proton-donor properties of the silanol groups of the silicon surface containing a thin (<3.0 nm) oxide layer suggest that for reactions to occur completely, one should either choose a reagent with $\sigma^*_i < \sigma_S^*$ or select the conditions of the additional activation of hydroxyls by raising the temperature or using proton acceptors as the catalysts of the exchange. For example, using triethylamine (TEA) in the synthesis of silica layers by processing the silicon surface alternately with $SiCl_4$ and H_2O vapors [9] allowed us to eliminate the latent period caused by the partial occupation of the surface by the silicon chloride groups.

Synthesis by the molecular superposition of silica, alumina, and tantalum oxide nanolayers at T = 403–603 K revealed the temperature effect on the layer growth parameter d_0, which is an increment in thickness per treatment cycle [9]. This dependence allows us to determine the mechanism of the formation of nanostructures and the conditions of layer growth. Using TEA allowed the elimination of the activation barrier, which led not only to higher yields of surface reactions, but also to a more than 100 K decrease in the layer growth temperature.

Of special interest is the preparation of silicon–nitrogen (in particular, nitride) structures by the ML technique because in the conventional processes, the synthesis of silicon nitride films occurs at high temperatures ($T > 1000$ K).

A distinction of the growth of Si_3N_4 nanolayers is a significant increase in the d_0 parameter and hence in the yield of the surface reactions with the synthesis temperature indicating that the reactions are activating [9, 10]. An X-ray photoelectron spectroscopy (XPS) analysis of the nitrogen states in the synthesized structures showed that nitrogen was mainly in the imide form at $T < 523$K and nitride form at higher temperatures [10]. Note that the nitride form can also form at reasonably low temperatures, as indicated by the XPS data, but its proportion is small, and pure silicon nitride is formed only at $T > 573$ K.

This is caused by the nucleophilic substitution

$$\equiv\text{Si–OH}, (\equiv\text{Si–O–Si}\equiv) + NH_3 \rightarrow \equiv\text{Si–}NH_2 + H_2O,$$

which starts at elevated temperatures ($T > 500$ K), leads to active imide groups, and activates the formation of the nitride form. Moreover, even a small fraction of the imide groups formed at lowered temperatures should suppress the protonation of hydroxyl groups due to the inductive effect. This, in turn, should lead to a low yield of the silicoimide form. Indeed, at elevated temperatures, the layer thickness d_0 for silicon nitride increased from 0.05 nm/cycle (T = 523 K) to 0.22 nm/cycle (T = 773 K. The ratio between these values and the Si–N distance ($l_{Si–N}$ = 0.27 nm) can serve as an estimate of the degree of surface filling during the reaction cycle. Since the concentration of hydroxyl groups on the surface decreased and layer growth was intensified as T increased, the yield increased at the first stage of the synthesis due to the nucleophilic substitution processes rather than the activation factor. This confirms the suggested scheme of the process.

REFERENCES

[1] V. B. Aleskovskii, *Vestn. Akad. Nauk SSSR*, No. 6, 48 (1975).
[2] V. B. Aleskovskii, *Zh. Prikl. Khim.* 55, 725 (1982).
[3] T. Suntola, *Mater. Sci. Rep.* 4, 261–312 (1989).
[4] Puurunen, R.L., *J. Appl. Phys.*, 2005, vol. 97, no. 12, 121301–121356.
[5] L. Colombo, T. Seidel, *Solid State Technol.*, No. 5, 111 (2004).
[6] Yu. K. Ezhovskii, *Russian Chemical Reviews*. V. 73, No 2, 195-204 (2004)
[7] Yu. K. Ezhovskii and P. M. Vainshtein, *Russ. J. Phys. Chem.* A71, 2009 (1997).
[8] Yu. K. Ezhovskii and P. M. Vainshtein,Russ. *J. Appl. Chem.* 71, 235 (1998).
[9] Yu. K. Ezhovskii, Khim. Fiz. 24 (4), 36 (2005).
[10] Yu. K. Ezhovskii, A. I. Klyuikov, *Zh. Prikl. Khim.* 63, 672 (1990).

DISPERSION FORCES IN NANOSCALE STRUCTURES METAL–DIELECTRIC–SEMICONDUCTOR

A. B. Fedortsov[1] *and V. A. Yurova**[,2]
[1]University of Mines, St. Petersburg, Russia
[2]The Bonch-Bruevich St. Petersburg State University of Telecommunications, St. Petersburg, Russia

ABSTRACT

The dispersion force pressure on the dielectric layer in metal–dielectric–semiconductor (MDS) structures is investigated using the Lifshitz theory at nonzero temperature. In this investigation the standard parameters of semiconductor devices with a thin dielectric layer are used. We consider the thickness of a layer decreasing from 80 nm to 3 nm. At the shortest thickness the dispersion pressure achieves 0.3 MPa. The results are compared with the predictions of nonrelativistic theory at small thicknesses (less than 5 nm).

Keywords: metal–dielectric–semiconductor (MDS) structures, silicon, palladium, aluminum, dispersion forces, Lifshitz theory

INTRODUCTION

Modern nanomechanisms and elements of microelectromechanical systems are made at micron and submicron levels. At this scales quantum effects and intermolecular interaction (dispersion forces) becomes commensurable on size with influence of typical electric forces. On developing nanoelectromechanical devices, it was found that one of their failure types can be caused by dispersion forces. At the same time, efforts are undertaken to develop nanoelectromechanical devices based on dispersion forces. This tendency motivated us to study dispersion force acting in nanoscale structures of modern electronics.

OBJECT OF RESEARCH

We choose the metal–dielectric–semiconductor (MDS) structure as the object to be investigated. These structures are widely used in developing integrated circuits as capacitors, field effect transistors, as charge coupled devices. Also this kind of structures may be used as a good and simple model for the first calculations of the dispersion pressure in MDS–structures. It is the first research of the dispersion pressure in real MDS–structure.

*Corresponding author: Valentina A. Yurova. Docent, the Bonch-Bruevich Saint - Petersburg State University of Telecommunications, Saint – Petersburg, Russia. E-mail: va-yurova@mail.ru.

One of the widely used MDS–structure is based on a silicon monocrystal substrate. It is manufacturing apply planar technology. A layer of a dioxide is formed on a substrate surface by oxidation of a surface by water steams. A metal layer is formed over it by thermal sprain methods.

The aims of our research are: to determine the dispersion pressure on the dielectric layer in the MDS–structure based on a silicon substrate; to estimate the influence of the dielectric thickness on the dispersion pressure.

CALCULATIONS

According to the Lifshitz theory the general formula for the dispersion pressure between two material layers separated with a gap of width a in thermal equilibrium at temperature T is:

$$P(a,T) = -\frac{k_B T}{8\pi a^3} \sum_{l=0}^{\infty}{}' \left(1 - \frac{\delta_{l0}}{2}\right) \int_0^{\infty} k_\perp dk_\perp K_l^{(0)} \sum_q \left[\frac{e^{2K_l^{(0)} a}}{r_q^{(0,1)}(i\xi_l, k_\perp) r_q^{(0,2)}(i\xi_l, k_\perp)} - 1 \right]^{-1}. \quad (1)$$

In this equation the notations are used: k_B is the Boltzmann constant; q is the magnetic and electric modes of the electromagnetic field; δ_{l0} is the Kronecker symbol; $k_\perp$ is the magnitude of the electromagnetic wave vector onto the plane of plates; r_q are the Fresnel reflection coefficients for two polarizations of the electromagnetic wave:

$$r_{TM}^{(SiO_2,n)}(i\xi_l, k_\perp) = \frac{\varepsilon_l^{(n)} K_l^{(SiO_2)} - \varepsilon_l^{(SiO_2)} K_l^{(n)}}{\varepsilon_l^{(n)} K_l^{(SiO_2)} + \varepsilon_l^{(SiO_2)} K_l^{(n)}}, \quad r_{TE}^{(SiO_2,n)}(i\xi_l, k_\perp) = \frac{K_l^{(SiO_2)} - K_l^{(n)}}{K_l^{(SiO_2)} + K_l^{(n)}},$$

where

$$K_l^{(SiO_2,n)} \equiv \sqrt{k_\perp^2 + \varepsilon_l^{(SiO_2,n)} \frac{\xi_l^2}{c^2}}.$$

The formula shows that the dispersion pressure depends on the thickness of the dielectric layer and the material's permittivities.

In our research we used the simplest models of the material permittivities and the experimental data of the refraction coefficient along the imaginary frequency axis. To determine the sensitive of the value of the dispersion pressure, we used two models of metal permittivity in the calculation: Drude and plasma. These models describe the electromagnetic and atomic properties of materials. The distinction of these models consist in the different account for energy dissipation in collisions of free conduction electrons with lattice ions and the account of the contribution of interband transitions.

RESULTS

Firstly, we researched the influence of the thickness of aluminum and palladium layer on the dispersion pressure in MDS–structure. The thickness of the dielectric layer was 80 nm. It has been established that the value of the dispersion pressure in the MDS–structure is practically independent on the thickness of metal layer. The type of permittivity model doesn't influence this value too.

It has been established that the dispersion pressure strongly depends on the dielectric thickness: when the thickness decreases from 80 nm to 40 nm, the dispersion pressure increases from 5 Pa to almost 70 Pa. The obtained dispersion pressures on the dielectric layer in the MDS–structure are in good agreement with the results of experiments on measuring dispersion forces between two plates separated vacuum gap [1].

Further decrease of the dielectric thickness leads to the increase of the value of the dispersion pressure by several orders of the magnitude. And when the thickness is 3 nm, the dispersion pressure is $3 \cdot 10^5$ Pa (fig.1).

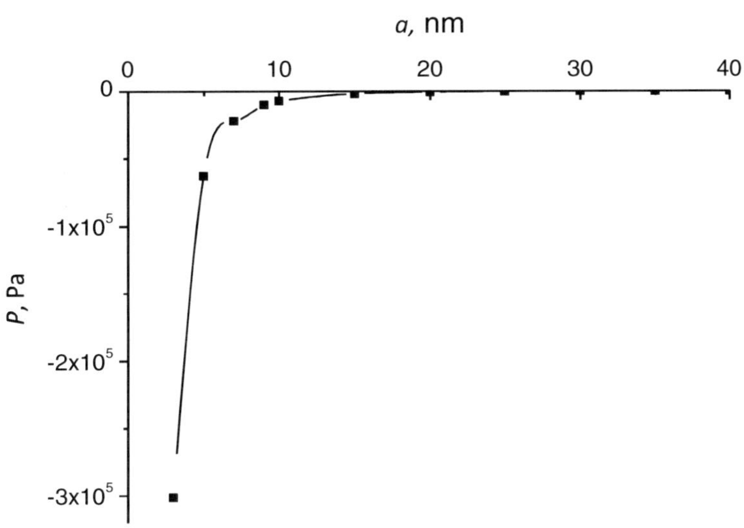

Figure 1. The dispersion pressure in the Al–SiO_2–Si metal-oxide-semiconductor structure as a function of thickness of the insulating SiO_2 layer.

Discussion

One nanometer is about three or five atomic layers. In this case we have to take into account the atomic structure of materials to describe permittivities and calculate the value of the dispersion pressure in MDS–structure with the dielectric thickness under 3 nm. It's the subject for our future research.

As the distance between conducting layers can be less than the characteristic wavelength of the absorption spectrum, we calculate the dispersion pressure in the nonrelativistic limit to.

In the nonrelativistic limit the formula (2) for the dispersion pressure assumes that the pressure's value inversely as the cube of the thickness of the dielectric layer [2]:

$$P(a) = -\frac{H}{6\pi a^3}. \tag{2}$$

In this formula H is the Hamaker constant:

$$H = \frac{3\hbar}{8\pi}\int_0^\infty d\xi \int_0^\infty y^2 dy \left[\frac{e^y}{r_{\text{Al}}(i\xi) r_{\text{Si}}(i\xi)} - 1\right]^{-1}.$$

For our MDS–structure the Hamaker constant is $1.61 \cdot 10^{-19}$ J in the case with using Drude model of aluminum permittivity. And it is $1.60 \cdot 10^{-19}$ J in the case with using plasma model of aluminum permittivity. So the result of calculations is independent of the model used to describe the aluminum permittivity. The values of the dispersion pressure are similar on relativistic and nonrelativistic limits. The difference between the results is less than 3 %.

Conclusion

It was theoretical investigated the dispersion force pressure on the dielectric layer in metal–dielectric–semiconductor (MDS) structures using the Lifshitz theory at nonzero temperature. In this investigation the standard parameters of semiconductor devices with a thin dielectric layer are used. The dispersion pressure increases from 5 Pa to almost 10^5 Pa, if the thickness of the dielectric layer decreases from 80 nm to 3 nm. This value does not depend on the aluminum or palladium layer thickness and the model of its permittivity. Thus the carried-out calculations showed that in metal–dielectric–semiconductor structures based on a silicon monocrystal substrate with aluminum or palladium metal layer the dispersion pressure, putting on the dielectric layer, amount to considerable value. Apparently, it makes sense to consider it when designing devices. Apparently, it makes sense to take this value into account when developed the electronically devices with MDS-structures.

References

[1] Chen F. Investigation of the Casimir force between metal and semiconductor test bodies / F. Chen, U. Mohideen, G.L. Klimchitskaya, V.M. Mostepanenko // *Physical Review* A, 72, (2005) 020101 – 4.

[2] Bressi G. Measurement of the Casimir force between parallel metallic surfaces / G. Bressi, G. Carugno, R. Onofrio, G. Ruoso // *Physical Review Letters*, 88, (2002) 041804– 4.

CLORINE-SILICATE PHOSPHOR FOR WHITE LIGHT-EMITTING DIODES

Maria V. Keskinova, Konstantin A. Ogurtsov, Maxim M. Sychov, Sergey P. Bogdanov, and Vadim V. Bakhmetyev
Department of Materials Science,
St. Petersburg State Institute of Technology (Technical University),
Saint Petersburg, Russia

The development of new effective phosphors is an important problem for the enhancement of light-emitting diode performances. Particularly, a promising type of phosphors useful for this application includes such compounds as $Ba_2Zn_3Si_3O_{11}:Eu^{2+}$, $Ba_3Si_6O_{12}N_2:Ce^{3+},Eu^{2+}$, $Sr_8[Si_4O_{12}]Cl_8:Eu^{2+}$, $Sr_2SiO_4:Eu^{2+}$ [1-4] generally featuring with the use of Eu^{3+} compounds in the reaction mixture for the synthesis followed by annealing in the reductive atmosphere of CO or H_2 to transform Eu^{3+} into Eu^{2+}.

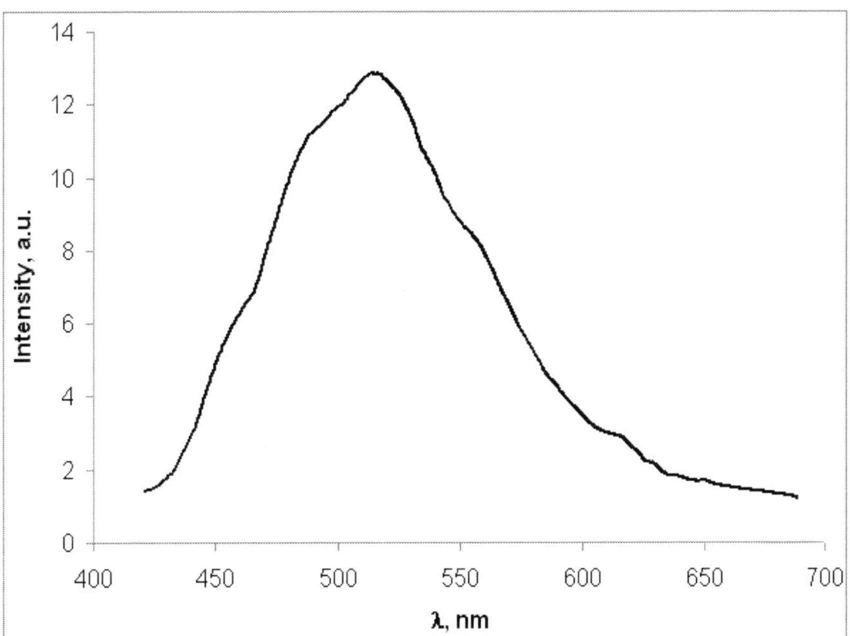

Figure 1. Spectrum of luminescence of chlorine-silicate phosphor.

The aim of this research work is the development of technology for synthesis of phosphors for white light-emitting diodes with improved characteristics. For this purpose, chlorine-silicate phosphors were synthesized using two methods, i.e., in the atmosphere of either CO or H_2+N_2. SiO_2, $SrCl_2 \cdot 6H_2O$, $Ca(OH)_2$, Eu_2O_3 preliminarily grinded in the mortar, sifted through the sieve and mixed during 3 hours were used as initial reagents.

The synthesis in CO atmosphere by annealing of the mixture in closed crucible under a layer of coal at 1030 °C resulted in obtaining samples with very low luminescence brightness. According to X-ray diffraction data, the poor performances of these phosphors were determined by the formation of $SrCO_3$ and $CaCO_3$ impurities in the phosphor.

The synthesis carried out in H_2:N_2 (5 : 95%) atmosphere at 700 °C allowed us to obtain carbonate-free chloric-silicate phosphors with a significantly improved brightness (about one order higher compared with the samples synthesized in CO atmosphere). The luminescence spectrum of thus prepared phosphors shown in Figure 1 is featured with the intensity peak in the green region (about 520 nm) and color coordinates x = 0,237 y = 0,379 (Figure 2). The obtained luminescence performances quite meet the requirements to efficient phosphors for the manufacture of white light-emitting diodes.

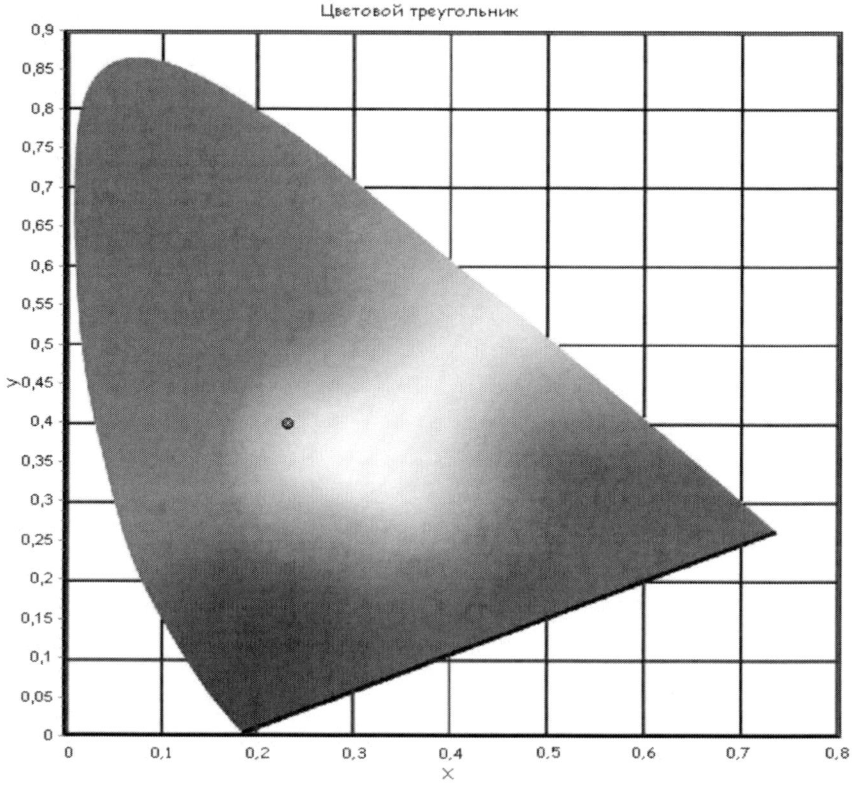

Figure 2. Color coordinates of chlorine-silicate phosphor.

References

[1] Donglei Wei, YanlinHuang, SunIlKim, YoungMoonYu, HyoJinSeo, *Materials letters,* 99, (2013) 122-124.

[2] Chenxia Li, Hong Chen, Youjie Hua, LanlanYu, Qingyun Jiang, Degang Deng, Shilong Zhao, Hongping Mab, ShiqingXu, *Optocs Communications*, 295, (2013) 129-133.

[3] Jiaguo Wang, Guobao Li, Shujian Tian, Fuhui Liao, Xiping Jing, *Material Research Bulletin,* 36, (2001) 2051-2057.
[4] Jee Hee Lee, Young Jin Kim, *Material Science and Engineering* B, 146, (2008) 99-102.

CHEMISORPTIONS FIELD EFFECT IN NANOCRYSTALLINE FILMS OF VANADIUM OXIDE

E. A. Tutov*[*,1], P. I. Kryukov[1,2], and V. P. Zlomanov[3]
[1]Voronezh State University of Architecture and Civil Engineering, Voronezh, Russia
[2]Voronezh State University, Voronezh, Russia
[3]Moscow State University, Moscow, Russia

ABSTRACT

Heterophase films of vanadium oxide on single-crystal silicon substrates have been synthesized by the sol–gel method from solutions of triethoxyvanadyl in methyl cellosolve. It has been found that the temperature of the semiconductor–metal phase transition in vanadium dioxide increases in a reducing atmosphere (ethanol vapor) and decreases in an oxidizing atmosphere (ozone).

Keywords: vanadium dioxide, semiconductor-metal phase transition, chemisorptions

INTRODUCTION

Fundamental and applied investigations of the semiconductor–metal phase transition in vanadium dioxide, which has first been described as far back as 1959, are still being pursued with unwaning interest. Despite the fairly large number of scientific publications, only thermistors classed among devices of engineering significance. One of the basic factors curtailing broad application of the effect is the large number of compounds in the vanadium – oxygen system and the ease with which comparatively weak actions can induce deviations from stoichiometry in the oxide phases of vanadium oxide, which leads to instabilities of their electrophysical characteristics.

On the other hand, this provides a possibility of controlling within fairly broad limits the main parameters of the phase transition, more specifically, the temperature, shape and width of the hysteresis loop, and the range of conductivity variation [1]. The effects produced by gas chemisorptions on the properties of vanadium dioxide have not been adequately studied, and their application in sensors is not, as a rule, directly related to the phase transition.

Using the chemisorptions-based field effect appears a promising approach to mediating the electron subsystem of surface states of a semiconductor, and this prompted to devote this chapter study of the effect exerted by chemisorptions of donor (ethanol vapor) and acceptor (ozone) gases on the parameters of the semiconductor–metal phase transition in vanadium dioxide films.

*E-mail: tutov_ea@mail.ru.

EXPERIMENTAL DETAILS

Following the accepted technique [2], films of vanadium oxide were synthesized from a solution of triethoxyvanadyl $VO(OEt)_3$ in methyl cellosolve $CH_3OCH_2CH_2OH$ by depositing it by centrifuging on a (001) oriented substrate of *n*-type single-crystal silicon. Next the sample was annealed in a muffle furnace in air at 500 °C to form vanadium pentoxide. The films thus obtained were homogeneous (the microstructure was studied with a JEOL-6380LV scanning electron microscope), with a good adhesion to silicon. Next the films were subjected to reduction annealing (for 10 – 30 min) in a flow of a 5 vol. % H_2 + 95% Ar mixture at 450 °C to produce lower oxides of vanadium. The phase analysis of the samples was conducted with X-ray diffractometer operated with filtered K_α radiation of cobalt. The crystallite size was inferred with the Debye – Scherer relation. The electrical dc and ac resistances of the $Ag/VO_x/Si$ structures (the ac measurements were conducted with an INSTEK, model 819 LCR meter) were studied at temperatures ranging from room level to 100 °C. The measurements were performed in a tubular heater in a microcompressor-produced flow of air saturated with ethyl alcohol vapor (making up about 5 vol. %). In another run of measurements, ozone was added to air (to a concentration on the order of 10 ppb). The rate of heating and cooling of a sample was approximately 1°C/min.

RESULTS AND DISCUSSION

The diffraction pattern of the film of higher vanadium oxide on silicon subjected to reduction annealing for 30 min demonstrates reflections of phases of the V_2O_5, V_6O_{13} and VO_2 oxides with crystallites 15–100 nm in size. The clearly pronounced orientation of all vanadium dioxides in the (001) plane engages attention. Studies of the behavior with temperature of the dc and ac resistance of a structure with a heterophase vanadium oxide film containing the VO_2(B) phase revealed stepwise variation of conductivity and a hysteresis loop in the 50–70 °C temperature interval. The width of the hysteresis loop for the phase transition in vanadium dioxide is about 15 °C. Chemisorptions of ethanol vapor drive the phase transition into the high temperature region by more than 10 degrees, with the hysteresis loop becoming noticeably broader (by approximately 4 °C). The total resistance of the structure increases. This manifestation of the chemisorptions field effect is typical of the interaction of a donor gas (reducer) with a *p*-type semiconductor. Vanadium dioxide residing in a reducing medium should maintain *n*-type conduction mediated by vanadium ions with a lower degree of oxidation (V^{+3}); its capacity to oxidation by oxygen of the air is, however, well known, and many authors suggest that VO_2 surface can accommodate the higher oxide V_2O_5. The hole conduction of vanadium dioxide may in this case be accounted for by the anion being present in excess of the stoichiometric ratio, at least in surface layers, and it is these layers that define adsorptive interaction.

The effect of ozone present in a moderate concentration (at a level of 10 ppb) on a phase transition is not so significant and acts in the opposite direction by lowering somewhat the transition temperature and film resistance on the whole.

The mechanism underlying the effect of chemisorptions of donor and acceptor gases on the parameters of the semiconductor–metal phase transition in vanadium dioxide suggests the following assumptions. Sorption of atmospheric oxygen on the surface of a film in *n*-type vanadium dioxide can, in principle, produce an inversion layer (with hole conduction) whose resistance in a reducing atmosphere should increase, exactly what we observe in experiment. Ozone intensifies the effect of oxygen to an insignificant extent only. The effect of chemisorptions of gases on the position and shape of the hysteresis loop observed in heterophase films may be mediated not only by processes occurring directly in vanadium dioxide but by reactions in the phases of other oxides (for instance, dehydration of vanadium pentoxide in ethanol vapor or oxidation-reduction of V_6O_{13}).

An influence of reversible chemisorptions of ethanol vapors on parameters of semiconductor – metal phase transition in VO_2-based film thermistor TP-68 is investigated by the measurements of its dc and ac conductivity and show the same results.

Conclusion

Perfectly ordered films of vanadium dioxide have been fabricated on single-crystal silicon substrate by the sol–gel method. X-ray diffraction analysis revealed the presence in the films of the V_2O_5, V_6O_{13} and VO_2(B) phases with crystallites 15–100 nm in size. Vanadium dioxide undergoes a reversible transition between states with a high and a low conductivity at a temperature of about 70°C, whose parameters can be controlled using gas chemisorptions. It has been found that chemisorptions of an acceptor gas (ozone) reduces the resistance of a structure with a vanadium dioxide film and the phase transition temperature, and that of a donor gas (ethanol vapor), raises it, a feature characteristic of a hole semiconductor.

References

[1] E.A. Tutov, V.P. Zlomanov. Effect of chemisorption of donor and acceptor gases on the semiconductor-metal phase transition in vanadium dioxide films. *Physics of the Solid State*, 2013. V. 55, № 11. P. 2351 - 2354.

[2] D.A. Vinichenko, V.P. Zlomanov, V.A. Vasil'ev, D.S.Seregin, O.Y. Berezina. Synthesis of vanadium dioxide films by a modified sol-gel process. *Inorganic Materials*, 2011. V. 47, Iss. 3. P. 279-284.

SYNTHESIS OF CADMIUM SELENIDE COLLOIDAL NANOPARTICLES STABILIZED WITH THIOGLYCOLIC ACID

D. S. Mazing and O. A. Aleksandrova
St. Petersburg Electrotechnical University "LETI", Russia

ABSTRACT

Colloidal nanoparticles of cadmium selenide with average size ranging from less than 1,5 nm to 2,2 nm were obtained in water medium using thioglycolic acid (TGA) acting as surfactant. Series of experiments was performed in order to reveal their characteristics and size dependence on technological parameters. Characterization methods include absorption and photoluminescence spectroscopies, X-ray diffraction analysis. Synthesized nanocrystals might be used for biomedical detection.

INTRODUCTION

Semiconductor colloidal quantum dots (QDs) have recently attracted much attention due to their tunable optical properties and technological flexibility. Owing to strong quantum confinement cadmium selenide nanoparticles exhibit luminescence in almost entire visible range depending on their size. One of the emerging fields for their application is biological and medical detecting [1]. In comparison with organic fluorophores semiconductor nanocrystals are much more stable to photodegradation. Moreover absorption band wideness and variety of available emission colors make it possible to excite plurality of QDs with different sizes by means of the same radiation source, thus allowing detecting signals from different targets simultaneously. Since hot – injection method was proposed [2] plenty of adaptations of colloidal quantum dots synthesis were developed aimed to reduce cost and hazards associated with usage of toxic and pyrophoric compounds. One of the natural choices of medium for QDs preparation might be polar solvents, for example water. Despite poorer crystallinity and therefore generally inferior optical properties as compared to nanoparticles synthesized in non-polar media these methods tend to be more environmentally benign and provide nanocrystals compatibility with biological tissues eliminating necessity for ligand exchange operations. Suitable precursors in this case may include water-soluble metal and chalcogenide salts as sources of metal and chalcogenide ions respectively while thiols might be used for surface passivation.

EXPERIMENT AND RESULTS

$NaBH_4$, $CdCl_2 \cdot 2.5H_2O$, elemental Se (powder), thioglycolic acid (TGA, 80%), KOH (6M), propanol-2 (anhydrous) were used without any further purification. Distilled water was used as solvent.

Typical synthesis procedure included following steps. Sodium hydroselenide solution was prepared by dissolving $NaBH_4$ and Se powder in distilled water in a glass vial. It resulted in vigorous reaction accompanied with intensive hydrogen gassing. The reaction might be described as follows:

$$4NaBH_4 + 2Se + 7H_2O \rightarrow 2NaHSe + Na_2B_4O_7 + 14H_2$$

To obtain cadmium precursor solution cadmium chloride hydrate was dissolved in distilled water and certain amount of TGA was added so that the mixture became turbid. By dropwise addition of KOH (6M) acidity of the solution was brought up to approximate value pH = 11 and it became clear.

In one implementation molar ratio $[Cd^{2+}]:[Se^{2-}]:[TGA]=1:0.25:1.2$ was used. NaHSe precursor solution was swiftly injected in cadmium containing solution at 45 °C which resulted in rapid nucleation indicated by color change. Temperature was gradually raised throughout the reaction and aliquots were taken consequently in following intervals: 1-1 – 0.5 min (50 °C), 1-2 – 6 min (51 °C), 1-3 – 22 min (58 °C), 1-4 – 60 min (63 °C), 1-5 – 90 min (64 °C), 1-6 – 125 min (95 °C). Resulting spectra are shown in the Figures 1 and 2.

In order to determine average size of the particles polynomial expression [3] was used:

$$a=\left(1.6122\cdot10^{-9}\right)\lambda^4-\left(2.6575\cdot10^{-6}\right)\lambda^3+\left(1.6242\cdot10^{-3}\right)\lambda^2-0.4277\lambda+41.57, \quad (1)$$

where a is average size of particles, λ is a wavelength of the first excitonic peak.

This expression gives approximate average size of QDs less than 1.6 nm. Absorption curves retain the first excitonic transition maximum prompting uniform growth and relatively stable monodispersity in the course of the reaction. Thus obtained nanoparticles exhibited broad band emission with the peak maximum in the range from 580 to 640 nm and featured large Stokes shift. It indicates that photoluminescence probably results from radiative recombination via trap states on the crystal surface. It is also notable that increasing temperature led to higher growth rate.

X – ray diffraction analysis was performed on one of the samples transformed into the form of powder through addition of propanol-2 with centrifugation. It showed that nanocrystals likely have hexagonal wurzite structure which is unusual for CdSe nanoparticles prepared in water solutions [4,5].

Another synthesis was implemented at higher injection temperature, so that precursors were mixed at 97 °C and this value maintained constant. Samples were consequently acquired in 0.5; 6; 25; 60 and 90 minutes after start of the reaction. Molar ratio was $[Cd^{2+}]:[Se^{2-}]:[TGA]=1:0.6:1.4$.

As can be seen from absorption spectra (Figure 3) comparatively larger nanocrystals were formed on nucleation stage and growth rate slowed down in approximately 25 minutes. In this case average size of the particles estimated from first excitonic peak wavelength values by means of expression (1) was less than 2.2 nm. Photoluminescence peak maximum shifts from 615 to 683 nm (Figure 4).

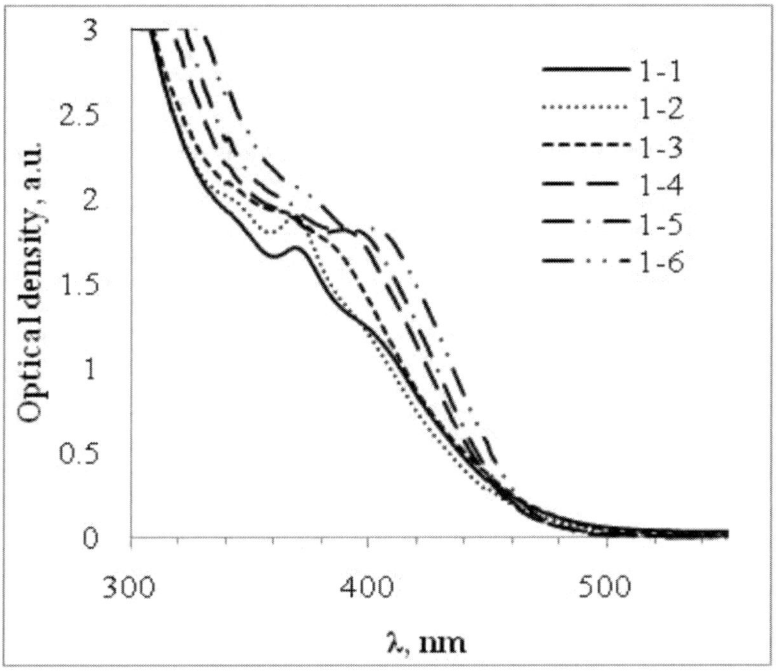

Figure 1. Absorption spectra of CdSe nanoparticles prepared with molar ratio $[Cd^{2+}]$: $[Se^{2-}]$:[TGA]=1:0.25:1.2 and injection temperature 45°C.

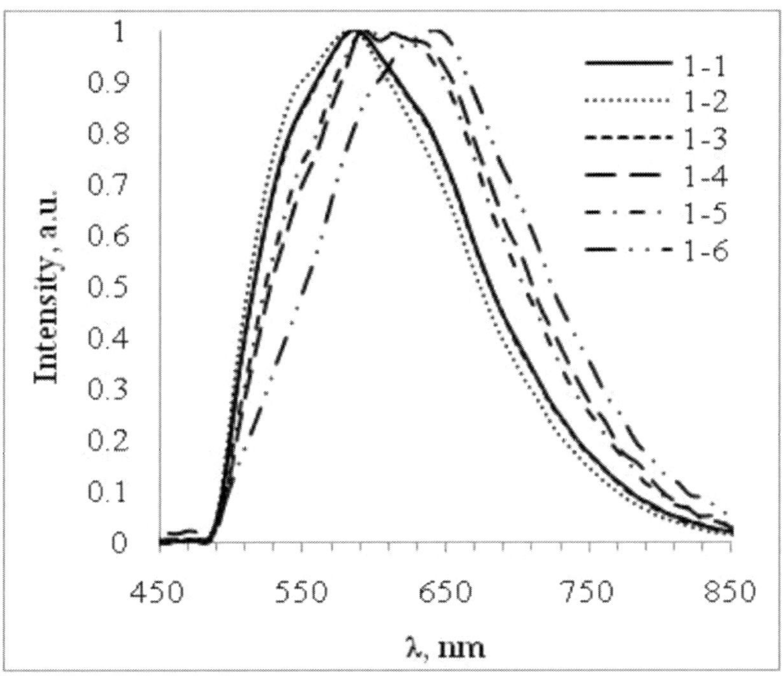

Figure 2. Photoluminescence spectra of CdSe nanoparticles prepared with molar ratio $[Cd^{2+}]$: $[Se^{2-}]$:[TGA]=1:0.25:1.2 and injection temperature 45°C.

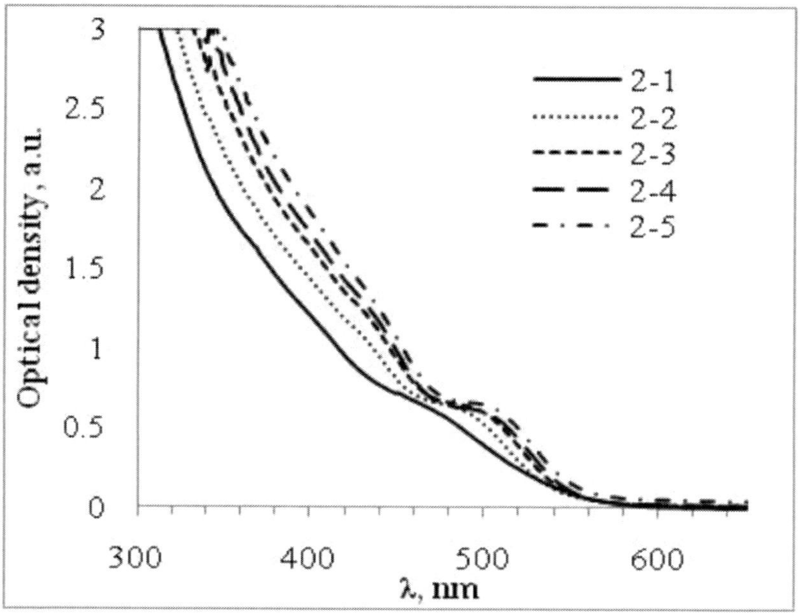

Figure 3. Absorption spectra of CdSe nanoparticles prepared at 97°C.

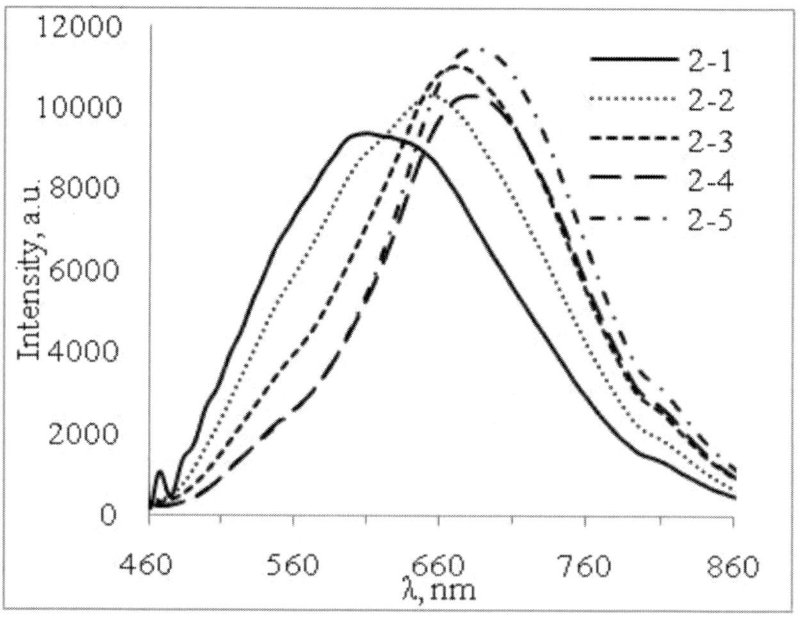

Figure 4. Photoluminescence spectra of CdSe nanoparticles prepared at 97°C.

Conclusion

In this chapter synthesis of CdSe nanocrystals was implemented through wet chemistry technique using thioglycolic acid acting as surfactant. Current method allows obtaining relatively small nanoparticles with tuning their average size by means of regulating temperature and component concentrations. Photoluminescence is characterized by broad peaks and large Stokes shift indicating that trap states play major role in emission process.

REFERENCES

[1] Walling M A, Novak J A and Shepard J R E 2009 *Int. J. Mol. Sci.* 10 441–491.
[2] Murray C B, Norris D J and Bawendi M G 1993 *J. Am. Chem. Soc.* 115 8706–15.
[3] Yu W W, Qu L H, Guo W Z and Peng X G 2003 *Chem. Mater.* 15 2854–60.
[4] Sun Q, Fu S, Dong T, Liu S and Huang C 2012 *Molecules* 17 8430–38.
[5] Sharma P K, Dutta R K, Liu C H, Pandey R and Pandey A C 2010 *Mat. Lett.* 64 1183–86.

INVESTIGATION OF THE OPTICAL PROPERTIES OF NANOPOROUS MEMBRANES BASED ON ALUMINA

L. B. Matyushkin and E. N. Muratova
St. Petersburg Electrotechnical University "LETI", Russia

ABSTRACT

Membranes of nanoporous alumina have been obtained using the electrochemical etching by varying technological regimes. The optical transmission measurements were performed on a spectrophotometer in the wavelength range of 190–1000 nm. It is possible to determine the average size and the dispersion of the pore diameter by UV-visible transmittance spectrum measuring.

INTRODUCTION

The electrochemically produced porous alumina (por-Al_2O_3) membranes have been widely used templates for the fabrication of ordered nanostructures [1, 2], owing to regular nanoscale structures and relatively easy and low-cost processing. Of considerable interest is the por-Al_2O_3 for the problems of biomedicine [3], for various types of sensors [4], in MEMS-devices [5]. A special attention is paid to the study of the optical properties of both the porous structures [6] and various composites based on them. Due to their structure, these membranes have potential in various optical device applications, such as optical filters and photonic crystals [7]. It is known that aluminum anodized in acidic electrolytes may occur at different rates, efficiency, and stability, in general, depend on the nature and concentration of the electrolyte, temperature and anodic current density [8].

It is important to be able to give a quick characteristic of the porous structure parameters, to evaluate the pore diameter, size distribution and order. Therefore, the aim of this chapter is to study the dependence of the transmission spectrum of the porous alumina membranes of their structural parameters.

EXPERIMENT AND RESULTS

In order to obtain a porous anodic alumina membrane electrochemical anodization of aluminum foil was conducted in potentiostatic mode in electrolytes based on aqueous solutions of phosphoric (H_3PO_4) and sulfuric (H_2SO_4) acids. Anodic alumina membranes contain many columnar pores with diameters are *about of 20-200 nm*. Geometric parameters of membranes obtained in different electrolytes are presented in Table 1.

The optical transmission measurements were performed on a PE-5400UV spectrophotometer (LLC "Ekohim") in the wavelength range of 190–1000 nm. The study of optical transmission spectra have been shown that for any membrane the transmission spectrum has the same form (Figure 1): the membrane starts to pass radiation with a certain wavelength.

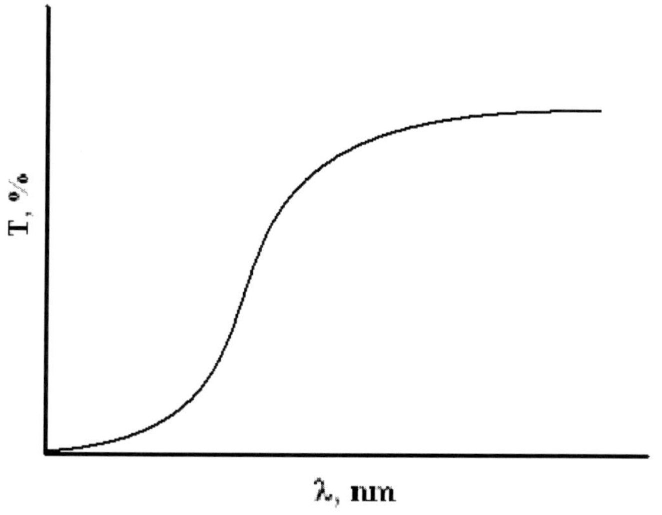

Figure 1. General view of the transmission spectra.

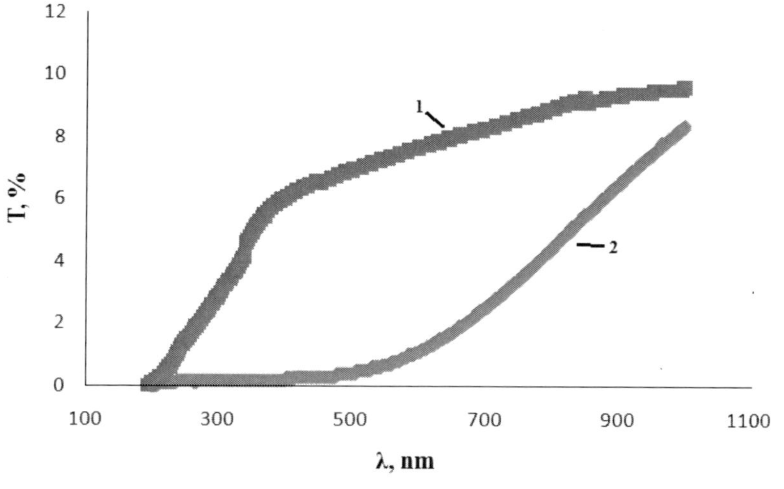

Figure 2. Transmission spectra of porous alumina membranes produced in presence of *1* – SO_4^{2-} and *2* – PO_4^{3-} anions.

A comparison of the transmission spectra for the different membranes (Figure 2) shows that the wavelength at which the membrane starts to pass radiation depends on the type of electrolyte in the presence of which it was produced.

It was found the following pattern: the smaller the pore diameter, the smaller the critical wavelength λ_c. Analys of experimental data shows some correlation between λ_c and type of porous structure due to electrolyte. Summary of the experimental date presented in Table 1.

Table 1. Influence of the electrolyte type on the geometric parameters porous layer and λ_c

Acid	Diameter of pore, nm	Thickness of membrane, nm	Concentration of pores, $1/\mu m^2$	λ_c, nm
sulfuric	20-30	12-18	350	210…280
phosphoric	80-100	4-7	50	450…750

Thus, the presented of experimental data generalization allows to determine the average size and the dispersion of the pore diameter by UV-visible transmission spectrum measuring.

Conclusion

Investigation has shown that the membrane starts to pass radiation with a certain wavelength, which depends on the type of electrolyte in the presence of which membrane was produced. Thus, it is possible to determine the average size and the dispersion of the pore diameter by UV-visible transmittance spectrum measuring. It could be put in the basis of the non-destructive express control technique for the determination of pore size in por-Al_2O_3.

References

[1] Eftekhari A. 2008 *Nanostructured Materials in Electrochemistry* (WILEY-VCH Verlag GmbH & Co. KGaA, Weinheim).

[2] Afanas'ev A V, Il'in, V A, Moshnikov V A, Sokolova E N and Spivak Yu M 2011 Synthesis of nanoporous and microporous structures by the electrochemical methods *Biotekhnosfera* no 1–2 pp 20-26.

[3] G Macias, L P Hernández-Eguía, J Ferré-Borrull 2013 Gold-Coated Ordered Nanoporous Anodic Alumina Bilayers for Future Label-Free Interferometric Biosensors *ACS Applied Materials & Interfaces* 5 (16) pp 8093–8098.

[4] Gracheva I E, Moshnikov V A, Karpova S S, et al. 2011 Net-like structured materials for gas sensors *Journal of Physics*: Conference Series 291 012017.

[5] Zimina T M, Muratova E N, Spivak Yu M., et al. 2012 Technologies of formation and application of nano-layers and nano-porous Al_2O_3 compositions for micro-and nanotech *Nano and Microsystems* no 12 pp 15-24.

[6] Akiyama T, Ishikawa Y, Hara K. 2013 Xylene sensor using double-layered thin film and Ni-deposited porous alumina *Sensors and Actuators B: Chemical.* vol 181 pp 348–352.

[7] Choi J, Luo Y, Wehrspohn R B et al. 2003 Perfect two-dimensional porous alumina photonic crystals with duplex oxide layers // *J. Appl. Phys.* 94, pp 4757.

[8] Muratova E N, Spivak Yu M, Moshnikov V A, Petrov D V, Shemukhin A A, Shimanova V V 2013 Influence of technological parameters of nanoporous Al2O3 layers' preparation on their structural characteristics *Glass Physics and Chemistry* vol 39 no 3 pp 320-328.

Enhancement of Vibro-Absorbing Properties for Poly(Vinyl Acetate) – Graphite Nanocomposites via Electron Beam Processing of Graphite Filler

***Sergey V. Mjakin*[1], *Maxim M. Sychov*[1], *Nadezhda B. Sheiko*[1], *Alexander G. Rodionov*[2], *and Inna V. Vasiljeva*[3]**

[1]St. Petersburg State Institute of Technology (Technical University), Russia
[2]JSC Plastpolymer, St. Petersburg, Russia
[3]Engineering Technology Center RADIANT, Ltd., St. Petersburg, Russia

Introduction

Polymer based composites involvong various dispersed, fibrous and layered fillers are one of the most promising types of vibro-protecting materials. Their vibro-damping efficiency is determined by viscoelastic properties of the polymer binder and interaction between the polymer binder macromolecules with the filler to afford a network of chemical bonds providing an additional absorption of mechanical vibrations. Therefore, the study and control of the filler surface functionality responsible for such interactions and formation of interfacial layers is a promising approach to the improvement of the resulting composite target performances.

A commonly used type of vibro-damping composites is based on poly(vinyl acetate) (PVA) as an available, plastic, thermally stable and non-toxic binder with various fillers including carbon materials, particularly graphite.

In this research we studied an approach to improvement of vibro-absorption properties of PVA based compositions due to the modification of graphite filler via electron beam (EB) processing earlier effectively used for the surface activation and enhancement of target properties for various materials [1].

Experimental

Graphite (GL-2, crystalline, foundry, ash content no more than 18%) filler was treated using the electron accelerator RTE-1V with the energy 900 keV, current 1 mA and absorbed dose in the range 25-600 kGy. The modified graphite was characterized by measuring the distribution of surface functional groups according to pKa values (using the adsorption of acid-base indicators as described in [1]) and incorporated in the amount 40 %wt. into the dispersions containing 40%wt. PVA dispersion (50%, GOST 18992-80, TU 2241-086-00203521-2001, average particle size 1 – 3 μm, density 1.17 g/cm^3 and dynamic viscosity 2.0 – 5.0 Pa·s), 7 %wt. dibutyl phthalate as a plasticizer and 13%wt. nepheline retardant.

Vibro-absorption properties of the obtained compositions were characterized by measuring the mechanical loss factor (k) according to Test Oberst procedure (TU 2243-038-

00203521-97) for bending vibrations of steel rods covered with layers of the studied composites.

RESULTS AND DISCUSSION

The obtained data indicate that EB of graphite filler results in a prominent increase in the mechanical loss factor of the studied composites. As shown in Figure 1, the achieved effect was found to correlate with the content of Brensted centers (hydroxyl groups). These centers appeared as a result of interaction with $H^{\bullet}$ and $OH^{\bullet}$ radicals (yielded by the EB induced radiolysis of adsorbed water molecules) can form a network of filler-binder and filler-filler donor-acceptor bonds thus enhancing the absorption of mechanical vibrations.

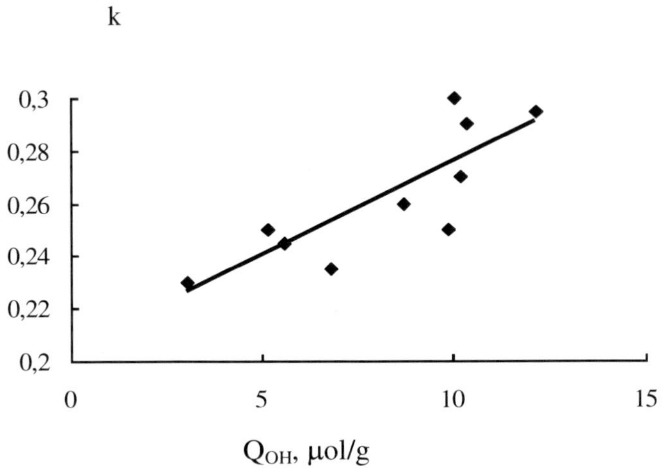

Figure 1. Mechanical loss factor of PVA based composites at 24°C as a function of hydroxyl groups content on the surface of modified graphite filler.

This conclusion is further confirmed by a drastically more significant growth of the mechanical loss factor at 35-38°C reaching about 2.5 times relating to the composites with non-modified graphite (table 1) compared with only ~20% growth at 24°C. These data suggest that a network of donor-acceptor bonds hampers the macromolecule segment motion at elevated temperatures.

Table 1. Mechanical loss factor at 35-38°C for PVA based composites with graphite fillers pretreated with accelerated electrons at various absorbed doses

Absorbed dose, kGy	Mechanical loss factor k
0	< 0.1
300	0.13
400	0.17
600	0.23

Conclusion

Generally the obtained results are promising for the prediction of vibro-absorption properties of composites on the basis of the analysis of the filler surface and obtaining vibro-protecting materials with improved properties (especially useful at elevated temperatures) due to the optimal selection and of fillers and specific functionalization of their surface.

Reference

[1] *Electron beam modification of solids: mechanisms, common features and promising applications*. Editors: Sergey V. Mjakin, Maxim M. Sychov, Inna V. Vasiljeva // Nova Science Publishers, Inc., 2009.

Band Model and Its Application to Describe the Structure of the Valence Band of Layered Semiconductors Based A2VB3VI

S. A. Nemov and N. M. Blagikh
St. Petersburg State Polytechnical University, St. Petersburg, Russia

The valence band of most semiconductors has a complex structure [1]. When the band structure is not well understood, it is traditionally used two-band model with two nonequivalent extremes, the picture for which the valence band is shown in Figure 1.

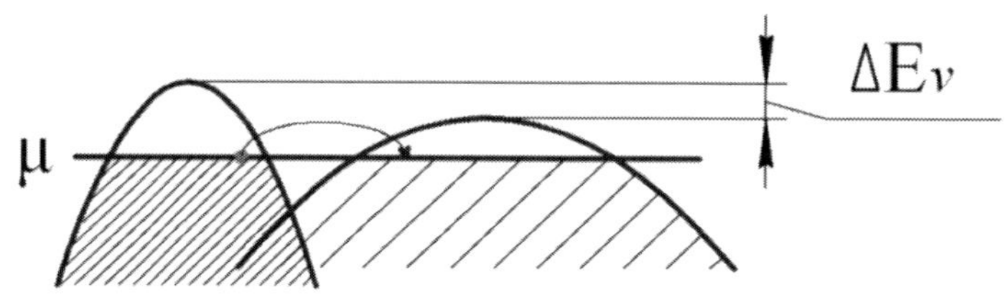

Figure 1. The structure of the valence band in the two-band model.

Four basic kinetic coefficients in the two-band model described by the following relations [2]:

$$\begin{cases} \sigma = \sigma_1 + \sigma_2 , \\ \alpha = \frac{\sigma_1}{\sigma}\alpha_1 + \frac{\sigma_2}{\sigma}\alpha_2, \\ R = \left(\frac{\sigma_1}{\sigma}\right)^2 \frac{A_1}{ep_1} + \left(\frac{\sigma_2}{\sigma}\right)^2 \frac{A_2}{ep_2} , \\ Q = \frac{\sigma_1}{\sigma}Q_1 + \frac{\sigma_2}{\sigma}Q_2 + \frac{\sigma_1\sigma_2}{\sigma}(\alpha_1 - \alpha_2)(u_1 - u_2) , \end{cases} \quad (1)$$

where the indices 1 and 2 – denote the corresponding partial kinetic coefficients of both extrema, and u_1 and u_2 – Hall mobility of holes in the first and second extremes.

For calculation of the kinetic coefficients in the two-band model is necessary to know the partial kinetic coefficients , Hall mobility , Hall factors , and the carrier concentration of the first and second extremum (σ_1, σ_2, α_1, α_2, Q_1, Q_2, u_1, u_2, A_1, A_2, p_1 and p_2), respectively. Temperature dependence of the partial kinetic coefficients in turn determines the dependence of the chemical potential $\mu(T)$ and the change in deposits dominant scattering mechanisms. Simplify the expression (1) can be expressed by the ratio of carrier concentration , mobility and Hall factor in the first and second extremes through $\eta = p_1/p_2$, $b = u_1/u_2$, $\gamma = A_1/A_2$, respectively. Within the two-band model, the system of equations (1) to calculate the temperature dependence of the chemical potential $\mu(T)$, usually supplement electroneutrality equation:

$p = p_1(T) + p_2(T) = \text{const.}$ (2)

Usually from experimental data on two-band formulas at low temperatures, making estimates of the parameters of the charge carriers of the second extremum (mass density of states m_{d2}, the width of the energy gap between the first and second extremum $\Delta E_v = \mu - \mu_2$, b, η, $\mu(T)$). However, the results thus do not allow parameters to describe the experimental temperature dependences.

In our opinion, this is due to the fact that not correctly taken into account the possible passage of charge carriers from the first to the second extremum without change of energy (elastic interband scattering). The first account of the interband scattering in the relaxation time τ (frequency transitions τ^{-1}) produced N.V. Kolomoets elements of the iron group transition metals [3]:

$$\begin{cases} \frac{1}{\tau} \sim |M_1|^2 g_1(\varepsilon), \varepsilon < \Delta E_v \\ \frac{1}{\tau} \sim |M_1|^2 g_1(\varepsilon) + |M_{12}|^2 g_2(\varepsilon), \varepsilon \geq \Delta E_v \end{cases}, \quad (3)$$

where M_1 and M_{12} - the transition matrix elements between the states within the first extremum and between the first and second extremes.

The complexity of calculating the frequencies of transitions represented by expressions (4) is that the dependence of the relaxation time on energy $\tau(\varepsilon)$ near $\mu \approx \Delta E_v$ becomes rapidly changing and Sommerfeld approach does not work. In this case, the partial kinetic coefficients to be calculated in terms of generalized Fermi integrals [2].

A similar calculation we had promising thermoelectric materials for hole-doped p-$PbSb_2Te_4$. It was found that when the concentration of holes about $1.6 \cdot 10^{20}$ cm^{-3} or less at low temperatures ($T \sim 100$ K) the contribution of the holes from the second extremum can be neglected and reliably enough to set the parameters of the holes of the first extremum (m_{d1}, μ, u_1, A_1).

With increasing temperature, the contribution of the second extremum increases and using the expressions for the kinetic coefficients in the two-band model (1) and the electroneutrality equation (2) is estimated m_{d2} and found other parameters that were used to describe the temperature dependence of the kinetic coefficients. Calculations of the temperature dependence of the formulas (1) using the expression for the partial kinetic coefficients of the generalized Fermi integrals (5) and taking into account the interband scattering similar to the model [3] with the dominant contribution of scattering by acoustic phonons in both extremes do not describe the experimental temperature dependences at temperatures above 200 K. significant difference between the calculated and experimental data at temperatures near room temperature, in our opinion, due to the need to account for the temperature dependence of the ratio of the Hall factors $\gamma(T)$.

Accounting $\gamma(T)$ in the interband scattering allowed in a two-band model to obtain a satisfactory agreement between calculated and experimental temperature dependences of the Hall coefficient and Seebeck in a wide temperature range from 100 to 350 K with the following parameters of the energy spectrum of holes at $T = 100$ K: $b \approx 4$, $m_{d1} \approx 0{,}5\ m_0$, $m_{d2} \approx 0{,}9\ m_0$, $\mu_1 \approx 0{,}22$ eV, $\Delta E_v\ (T) \approx 0{,}23\text{-}0{,}04 \cdot (T/100 - 1)$ eV (see. Fig. 2).

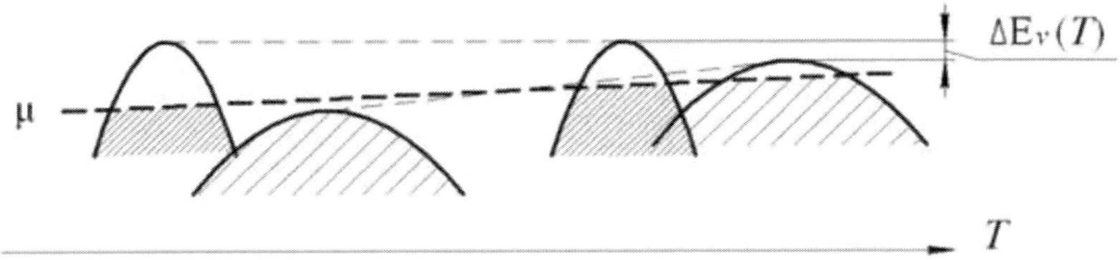

Figure 2. Pattern of variation of the energy spectrum and the position of the chemical potential with temperature $PbSb_2Te_4$.

References

[1] Zhitinskaya M.K., Nemov S.A., Ivanova L.D. *Physics of the solid state*, 2002, Vol. 44, Iss. 1, pp. 42-47.

[2] Аскеров Б.М. *Кинетические эффекты в полупроводниках* Л.: «Наука», 1970, 303 с.

[3] Коломоец Н.В. ФТТ, 1966, Т.8, вып. 4, с.997-1003.

EFFECT OF EXPLOSION TREATMENT OF ZNS ON THE LUMINESCENCE PROPERTIES OF AN OBTAINED PHOSPHOR

K. Ogurtcov[1], M. Sychov[1], V. Bakhmetyev[1], V. Lebedev[2], A. Sokolov[2] and V. Trunov[2]

[1]St. Petersburg State Institute of Technology (Technical University), St. Petersburg, Russia

[2]Petersburg Nuclear Physics Institute, Gatchina, Russia

INTRODUCTION

Increasing the brightness of the luminescence in system ZnS-Cu_2S is achieved for example by varying the concentration of the activator, however, there is an optimal, its concentration, determined in particular its solubility through the phosphor. Therefore, to increase the solubility of the activator (ie, copper) in zinc sulfide was invited to pre-treated ZnS in different ways to create additional defects in it, which would facilitate the diffusion of copper during the synthesis of the phosphor, thereby increasing the number of luminescent centers. In other words, by pretreatment ZnS create it additional nano defects that allow in the high-temperature synthesis of phosphors to create in its basis more and well-dispersed nanosized luminescence centers. Also, this treatment approach can be applied not only to the phosphor industry, but also in other areas where they have supersaturated solid solutions.

There are papers studying the treatment of zinc sulfide with electrons, as well as to study the impact of the explosion at the pure, undoped ZnS, and synthesis by the explosion of zinc sulfide and alkali halide phosphors [1-5].

MATERIALS AND METHODS

ZnS has been modified by explosion in a copper ampoule. ZnS was placed in a copper ampoule with a wall thickness of 3 mm, top and bottom screw cap, as well as to mitigate the explosion of a copper ampoule was placed in a steel tube, 6 mm thick. Then, the outside steel ampoule turned around explosive (RDX) and undermined. After which the ampoule saw, extracted ZnS and sieved through a 40 microns.

Further, with treated and untreated zinc sulfide prepared mixture containing various concentrations of copper. Synthesis of phosphors was carried out in corundum crucibles under a layer of activated carbon at a temperature of 950°C for 90 minutes. Using the synthesized phosphors were fabricated electroluminescent light sources AC.

Electroluminescent and photoluminescent brightness were measured with IL 1700 radiometer. Spectra of luminescence measured with spectrofluorimetre AvaSpec 3648 at room temperature. Measurement of diffuse reflection spectra were performed on a spectrophotometer brands SPECORD M-200 in the wavelength range 200-900 nm. X-ray analysis was performed by X-ray diffractometer «Rigaku Ultima IV» using Cu Kα-radiation and detector D/teX Ultra. The uniform distribution of activator was investigated by small-angle neutron scattering diffractometer "Yellow submarine".

RESULTS AND DISCUSSION

The fact that pre-treatment of zinc sulfide leads to the formation of defects in its structure is indirectly confirmed by diffuse reflectance spectra of the initial and treated by explosion of ZnS. From the data obtained shows that after treatment zinc sulfide darkening in the visible region (10-15%), and the band gap has not changed. This darkening is probably due to the release phase Zn^0 on the surface of ZnS, which confirms the hypothesis about the formation of vacancies to facilitate the introduction of copper.

X-ray analysis shows that treatment by explosion zinc sulfide decreases the value of the lattice constant (from 5.4102 Å to 5.4069 Å). After synthesis of phosphors on the contrary there is an increase of the lattice constant, which may be a consequence of incorporation into the structure of zinc sulfide larger amounts of copper, having a greater ionic radius than that of zinc (ionic radius of Zn^{2+} - 0.74 A, ionic radius of Cu^+ - 0.77 A). All the samples take the form of the cubic modification of ZnS (the zinc-blende structure, space group 216). It was also determined that increasing copper concentration from 0 to 0.15 wt%. an increase in the crystallite size of the synthesized phosphor (from 743 to 1006 A).

The pretreatment of zinc sulfide is likely to lead to the formation of additional defects (vacancies of zinc and sulfur), which facilitates the diffusion and incorporation of the activator and co-activator in the zinc sulfide lattice during the synthesis of phosphor. Increasing the number of "green" luminescence centers demonstrates the shift of the spectrum to longer wavelengths, Figure 1.

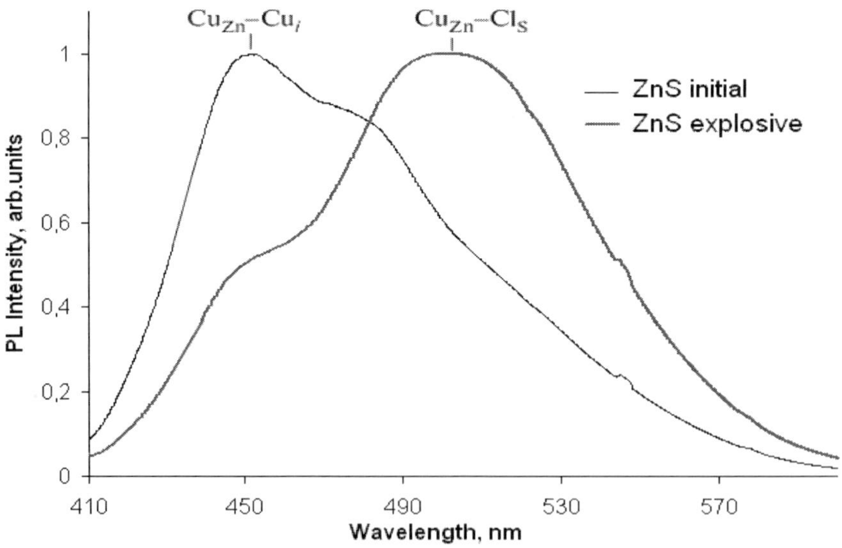

Figure 1. PL spectrum of the ZnS:Cu,Cl (0.3 wt.% Cu) phosphors with different ZnS.

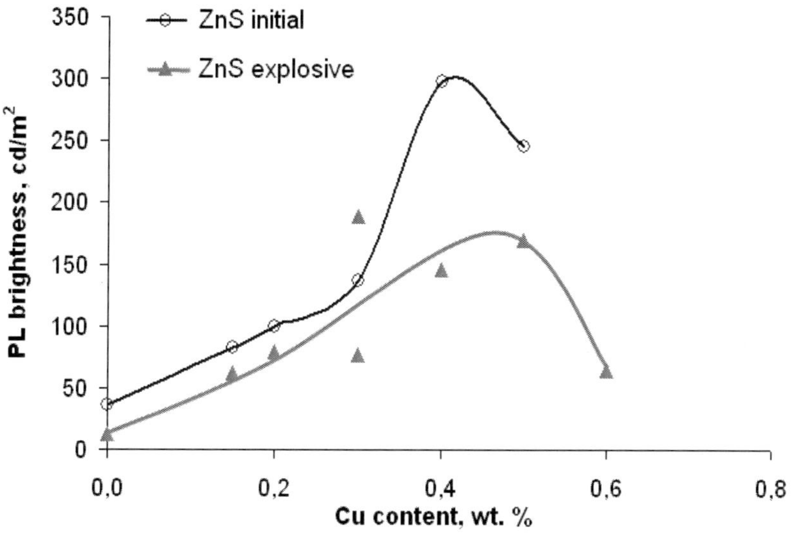

Figure 2. Brightness of the photoluminescence ZnS:Cu,Cl phosphors with different ZnS.

Such a large increase in the defect zinc sulfide after explosion treatment, which are supported by data of neutron diffraction pattern, leads to an increase in the number without radiative centers, which explains the lower brightness of the luminescence in comparison with other samples, Figure 2. Whereas pretreatment of ZnS by explosion shifts the optimum concentration of copper in a big way and is equal to 0.5 wt.%.

CONCLUSION

It is shown that the pretreatment of zinc sulfide by explosion induces shift of the luminescence spectrum to longer wavelengths due to the more homogeneous distribution of the activator in the matrix of the phosphor. This effect is attained due to the formation of extra structural nano-defects in ZnS during the treatment stage; the treatment induced defects accelerate diffusion and incorporation of the activator and co-activator into the structure during the high temperature stage of phosphor synthesis. In the case of the shockwave treatment of zinc sulfide, the generation of defects is not compensated by heat treatment during synthesis of the phosphor, and a positive effect of enhancing the luminescence brightness is not attained. Also, there is a problem of reproducibility of results.

REFERENCES

[1] Effect of zinc sulfide E-beam processing on EL phosphor properties / M.M. Sychov, V.V. Bakhmetiev, E.V. Komarov, et all. // *Proc. of ADT-04. – Minsk*, 2004 – Pp. 151 - 154.

[2] S.S. Batsanov, A.I. Lapshin / Action of the explosion on matter. Photoluminescence of ZnS after dynamic compression // *Combust.* - 1967 – V.3 - Pp. 441 - 448.

[3] A.I. Lapshin, T.P. Lazarenko, S.S. Batsanov. On the possibility of activation of crystal zinc sulfide type by explosive loading // *Journal of Applied Spectroscopy*. - 1971 - V. 14, Issue 6 - Pp. 1020 - 1026.

[4] A.I. Lapshin, V.V. Kournikova. Some features of alkali halide phosphors doped with Eu^{2+} by explosive loading // *Journal of Applied Spectroscopy* – 1978, Volume 28, Issue 1 - Pp. 95 - 100.

[5] Patent Japan 2002-153747. *Production of high pressure phase substance* / Ito Kenji, Hatake Tomoaki, Sakakibara Ikuo, Takahashi Katsuhiko.

SORPTION PROPERTIES OF COMPOSITE MATERIALS BASED ON CALCIUM HYDROXYAPATITE

D. L. Goloshchapov[1,2], A. S. Len'shin[1], E. A. Tutov*[,2], Yu. A. Ponomarev[2], and E. V. Maraeva[3]
[1]Voronezh State University, Voronezh, Russia
[2]Voronezh State University of Architecture and Civil Engineering, Voronezh, Russia
[3]St. Petersburg Electrotechnical University "LETI" Petersburg

ABSTRACT

Sorption and sensing properties of hydroxyapatite based materials was estimated by TEM, BET and dielectric loss in vapor with different values of relative humidity. It was shown that hydroxyapatite that was obtained by using hen's eggshell has nanocrystalline morphology with large surface area (55,4 ± 0,9 m^2/g) and it's dielectric loss have a specific feature with increasing of relative humidity.

INTRODUCTION

Nanosized materials with high sorption characteristics are often used as a part of different sensors and sensing device. There are several special demands for such materials as stability and durability of life. The estimation of different vapor gas content could be implementing with the usage of calcium phosphates, for example, hydroxyapatite (HAP). Because of its dielectric properties, stability on atmosphere, inactivity in interaction with various gas vapors and stability under heat treatment up to ~700 ^{0}C [1] this material could be used in different gas sensing device.

MATERIALS AND METHODS

Hydroxyapatite was prepared by wet precipitation rout that have been described in [1], with the usage of hen's eggshell as a calcium source. All materials were investigated by X-ray diffraction (XRD), transmission electron microscopy (TEM), Brunauer-Emmett-Teller surface area analysis (BET) by thermal desorption of nitrogen and impedance analysis.

For all type of investigation techniques samples was annealed at 400 ^{0}C, to exclude H_2O. Determination of phase analysis was provided on Thermo ARL X-TRA «Thermo Techno» with copper anode. For TEM analysis all samples have been dispersed by ultrasound procedure (44 kHz) and was applied to carbon grid, that was placed to the microscope holder. Sorption activity of HAP powders was calculated with Sorby-MS apparatus by the BET

*E-mail: tutov_ea@mail.ru.

theory. Dielectric characteristics of HAP was determined on LCR-meter GW; LCR-819. For this technique the samples was pressed with tin foil under high pressure (300 bar). The end thickness of such parallel-plate capacitor was ~1 mm. Dielectric measurements were provided in environment with various values of relative humidity (p/p_s) - 33, 65, 76, 85, 100%.

DISCUSSION

The data of X-ray diffraction show that all samples are calcium hydroxyapatite. Comparison between microcrystalline and obtained HAP samples revealed that the synthesized materials are nanocrystalline. Calculations by using Debye-Scherer's equation show that average crystal size is ~ 35 nm. Typical TEM images of annealed and dispersed samples are provided on Figure 1. It follows that HAP nanocrystals have shaft form with diameter ~ 20 nm and length ~ 50 nm.

Figure 1. TEM of hydroxyapatite samples.

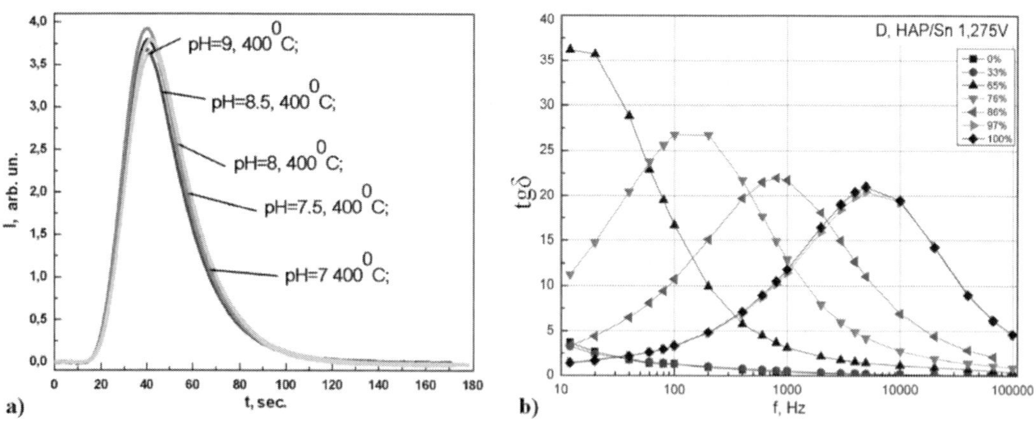

Figure 2. a) BET curves of nitrogen desorption; b) Dependence tgδ vs frequency for HAP samples at different values of relative humidity p/p_s.

Microdiffraction provided on Figure 1 contains the reflections that are referenced only to calcium hydroxyapatite. This fact, in addition to XRD, proved the single-phase state of all materials that was obtained in this chapter. With magnification to nanoscale range porous structure of HAP nanocrystals were discovered. This unique morphology was related to the nature of the samples and its obtaining route.

Sorption capability of hydroxyapatite powders was investigated by thermal desorption of nitrogen by BET theory. The desorption curves (Figure 2, a) show the presence of high surface area in the samples that were obtained under various conditions. The specific surface area (S_{sp}) calculated by BET theory is. 55,4 ± 0,9 m^2/g, that is much more higher than knowing value from literature source.

The data from dielectric loss (tgδ) of HAP show (Figure 2, b), that in frequency range 10Hz - 100 kHz and with relative humidity changes $0\% \leq p/p_s \leq 100\%$ the conductivity in all hydroxyapatite samples is determined by absorbed water.

In relative humidity range $0\% \leq p/p_s \leq 33\%$ and frequency range 10 Hz - 100 kHz the dielectric permeability of HAP ($\varepsilon \approx 8$) doesn't have any dependence from p/p_s. The special feature of dielectric loss tangent in HAP samples are that in relative humidity diapason $65\% \leq p/p_s \leq 100\%$ maximum of wide band (Figure 2, b) has a shift to the high frequencies range with the increasing p/p_s value.

CONCLUSION

An explanation of the impedance data could be provided with the suggestion of boundary charge occurring that are induced by external alternating electric field. This charges are placed on the boundary between conductive and isolators layers. In this case water/hydroxyapatite boundary could be observed as dipoles which direction of moments changes with the change of external field. The dielectric loss increase when the oscillation period of electric field strength is matching with the relaxation time of surface polarization. The received results prove one of the available applications of HAP materials as part of different sensors and sensing device.

REFERENCE

[1] Synthesis of nanocrystalline hydroxyapatite by precipitation using hen's eggshell. D.L. Goloshchapov, V.M. Kashkarov, N.A. Rumyantseva, P.V. Seredin, A.S. Lenshin, B.L. Agapov, E.P. Domashevskaya. *Ceramics International.* 2013. V. 39. Iss.4. P. 4539–4549.

INFLUENCE OF NANODIMENSIONAL EFFECTS ON ELECTRIC ADHESION IN ANODIC BONDING MANUFACTURING OF COMPOSITE SEALS

N. S. Pshchelko[1], V. A. Moshnikov[2], M. P. Sevryugina[1], and S. A. Moshkalev[3]

[1]University of Mines, St. Petersburg, Russia
[2]St. Petersburg State Electrotechnical University "LETI", Russia
[3]University of Campinas, Brazil

ABSTRACT

The appearance of great pulling electric fields at anodic bonding process with a conductor surface to dielectric turns out to be possible due to the interlayer polarization developing in dielectric under the action of electric voltage. This results in a negative charge accumulation in a layer of small thickness beside anode. Thus applied electric voltage is distributed not through the whole thickness of a dielectric, but in fact is applied to a narrow area of the three-dimensional charge beside anode. Arising strong electric fields force the connected materials to unite. The physical and mathematical models for force characteristics of non-uniform electric field are developed. Research of a charge intermittence influence on intensity of an electrostatic field has shown a considerable difference on small distances to the charged surface between carried out and traditional ways of calculation.

Keywords: electric field, discrete charge, electric adhesion

Actually processes of object's gluing, sticking and permanent or temporary fixing of objects is required in one form or another in almost all areas of production. For the design of new products and the introduction of advanced technologies not only the development of new materials with desired properties is relevant, but also the problem of their connection - gluing, welding, and other methods of permanent or temporary fixing. The compounds of materials that combine a variety of mechanical and electrical properties are required in almost all areas of production. Specifically, the problem of connection of the metals and other conductive materials, including semiconductors, with dielectrics is of particular interest.

The increase of surface's adhesion is the main task to be solved to obtain a reliable connection between two surfaces. There are two major factors that can increase the adhesion:

1. Diffusion. Strengthening of the heterogeneous bodies contact points may be provided by diffusion of atoms and molecules both of the two bodies, resulting in a blurring of the interface.
2. The formation of electrical double layer on the contact.

The present work is devoted to assessing the influence of electric charge distribution's discreteness in the electric double layer on the characteristics of the electric field and force characteristics of the electric adhesion contact. The calculation results show that the charge discreteness accounting leads to significant amendments in the direction of increasing of the field strength and the ponderomotive force acting on the charge at a relatively small distance in comparison with the heterogeneously charged plane.

Thus, electric fields can be significantly higher than the values calculated by the classical formulas at small distances to the charged plane. Given that the field is heterogeneous, it may have a retractor force action even on electrically neutral objects due to the gradient forces, for example, atoms sprayed on the substrate [1], and may have other manifestations [2].

REFERENCES

[1] Afanasev V. P., Chigirev D.A., Pshchelko N.S., Sidorova N.P. Influence of constant electric field on processes of sedimentation of thin metal films of platinum by a method of ionic-plasma dispersion // *News of HIGH SCHOOLS of Russia. Radio electronics*. 6, (2010), 59-65.

[2] Pshchelko N.S., Use of electric field for increase of adhesion of electrospending films to dielectric substrates at vacuum deposition // *Vacuum technics and technology*, 1, (2010) 31-36.

Formation of Titanium Oxide Semiconductor Structures by the Local Anodic Oxidation

A. I. Maksimov*[1]*, V. A. Moshnikov*[1]*, N. S. Pshchelko*[2]*,
***A. V. Startseva*[1]*, G. Suchanek*[1]*, and S. A. Moshkalev*[3]**
[1]St. Petersburg State Electrotechnical University "LETI", Russia
[2]University of Mines, St. Petersburg, Russia
[3]University of Campinas, Brazil

Abstract

In this paper we investigate the process of the titanium thin films local anodic oxidation. It is shown that the configuration of the formed oxide depends on the value and duration of the applied voltage, the ambient humidity, the pressing force of the probe to the sample surface, and the parameters of the probe used in the oxidation process. According to the research, it was found that the threshold voltage of the process beginning of titanium local oxidation in this case is 6 V. The applied voltage value dependence of the oxide thickness is linear, and the kinetics of the process for large times obeys Cabrera-Mott model.

Keywords: scanning probe microscopy, local anodic oxidation method

The miniaturization of electronic devices, an increase of recording density in data storage/read device, development of new functional biochemical sensors, creating micro-chemomechanical system to substitute silicon integrated microcircuits require a search for new materials and methods for forming nanoscale structures [1, 2].

Principles of scanning probe microscopy can be used to analyze, to characterize and to modify the surfaces. The creation of active microelements on the basis of metal structures for gas-sensitive sensors and obtaining the resistive structures based on the phases of variable composition with the hysteresis curve on the current-voltage characteristic [3] are the promising directions of using local anodic oxidation method (lithography with a scanning probe microscope).

This requires a detailed understanding of the processes occurring at the local anodic oxidation: geometric dimensions and configuration of the formed oxide structures depend on the applied voltage, the environment, and the parameters of the used probes and many other factors.

The aim of this work is to identify regularities and features of local anodic oxidation process for managed forming of thin nanoscale structures based on titanium oxide.

The influence of various parameters on the process of titanium thin films local anodic oxidation is studied and it's obtained the following results:

1. The threshold voltage of the process beginning of titanium local probe oxidation is about 6 V.

2. The oxide film growth kinetics is determined by the Mott-Cabrera mechanism. The applied voltage value dependence of the oxide height is linear. The oxide height depends on the time of voltage application logarithmically.
3. The pressing force of the probe to the sample surface dependence of oxide height is approximated by two linear sections: small portions of the cantilever bending at low clamping forces and a portion of the hard contact at high clamping forces. It is recommended to maintain low contact pressure of the probe in terms of increasing the height of the oxide.
4. The shape of water meniscus between the tip and the sample surface effects on the formed oxide geometry. In turn, humidity environment, the pressing force of the probe to the sample surface, the probe parameters used in the oxidation process and the wettability parameters of the sample surface and probe materials directly affect the shape of the meniscus.

REFERENCES

[1] Mackay A. L. Quasicrystals turn to the sixth-dimension//*Nature*., 344, (1990) 21.

[2] Лорд Э., Маккей А., Ранганатан С. Новая геометрия для новых материалов. М.: Физматлит, (2010), 264.

[3] Strukov D.B., Snider G.S., Stewart D.R., Williams R. S. The missing memristor found. – *Nature letters*, 453, (2008), 80–83.

PHASE TRANSITION IN NANOCRYSTAL VO2 THIN FILMS

A. V. Ilinskiy[1], O. E. Kvashenkina[2], and E. B. Shadrin[1*]
[1]Ioffe Physical Technical Institute of the Russian Academy of Sciences, Russia
[2]St. Petersburg State Polytechnical University, Russia

ABSTRACT

A complex concept concerning the multistage process of thermal semiconductor–metal phase transition in vanadium dioxide is discussed. According to the concept, the phase transformation proceeds in three stages: two stages occurring in a system of strongly correlated electrons and one stage in lattice system which increases of the crystal lattice symmetry by phase transformation. This concept generates a possibility of creation of a wide set of new applied devices

Keywords: vanadium dioxide, phase transition, strong correlations

The purpose of offered article is the description a complex concept concerning the multistage process of thermal semiconductor–metal phase transition in vanadium dioxide.

As it is known from X-ray structural analysis [1], the unit cell of the lattice of the tetragonal VO_2 phase contains two vanadium atoms and four oxygen atoms (Figure 1).

The valence bonding theory says that, in such a unit cell, due to $3d^2 4s^1 4p^3$ hybridization, each vanadium atom forms six σ-bonds with six oxygen atoms, and each of them has three sp^2 hybridized orbitals. All vanadium atoms are at the centers of oxygen octahedrons. The remaining three nonbonding $3d$ orbitals (d_{xz}, d_{yz}, d_{x2-y2}) of the vanadium atom, with one electron belonging to them, do not participate in the formation of the skeleton of σ-bonds of the oxygen octahedron. Except for σ-bonds, in the oxygen octahedron, there are saturated π-bonds generated from orbitals of vanadium (V^{4+}) ions and $2p_z$ orbitals of oxygen ions. According to the theory of molecular orbitals, these π-bonds form π and π* electron energy levels. In the crystal, these levels broaden into π and π* energy bands. The d_{x2-y2}-orbital has zero overlapping with all of the ligand orbitals.

It is nonbonding and has the lowest energy among the orbitals that do not have connections with the V^{4+} ion. The d_{x2-y2}-orbital is not involved in the formation of the octahedron and takes a one electron remaining to the all d_{xz}, d_{yz} and d_{x2-y2} orbitals. According to X-ray structural data [1], the oxygen octahedrons are arranged so that the tetragonal phase contains quasi one dimensional (1D) chains of vanadium ions located at the centers of oxygen octahedrons; the chains are parallel to the rutile C_R axis (Figure 1). Thus the orbital containing one electron for each ion with branches arranged along the x and y axes can be responsible for a nonzero overlap integral only with a similar orbital of V-ions in neighboring octahedrons arranged along the C_R axis.

*Corresponding author: Prof. E.B.Shadrin. Professor of Ioffe Institute, St. Petersburg, Russia. Tel.: + 79095798026. E-mail: shard.solid@mail.ioffe.ru.

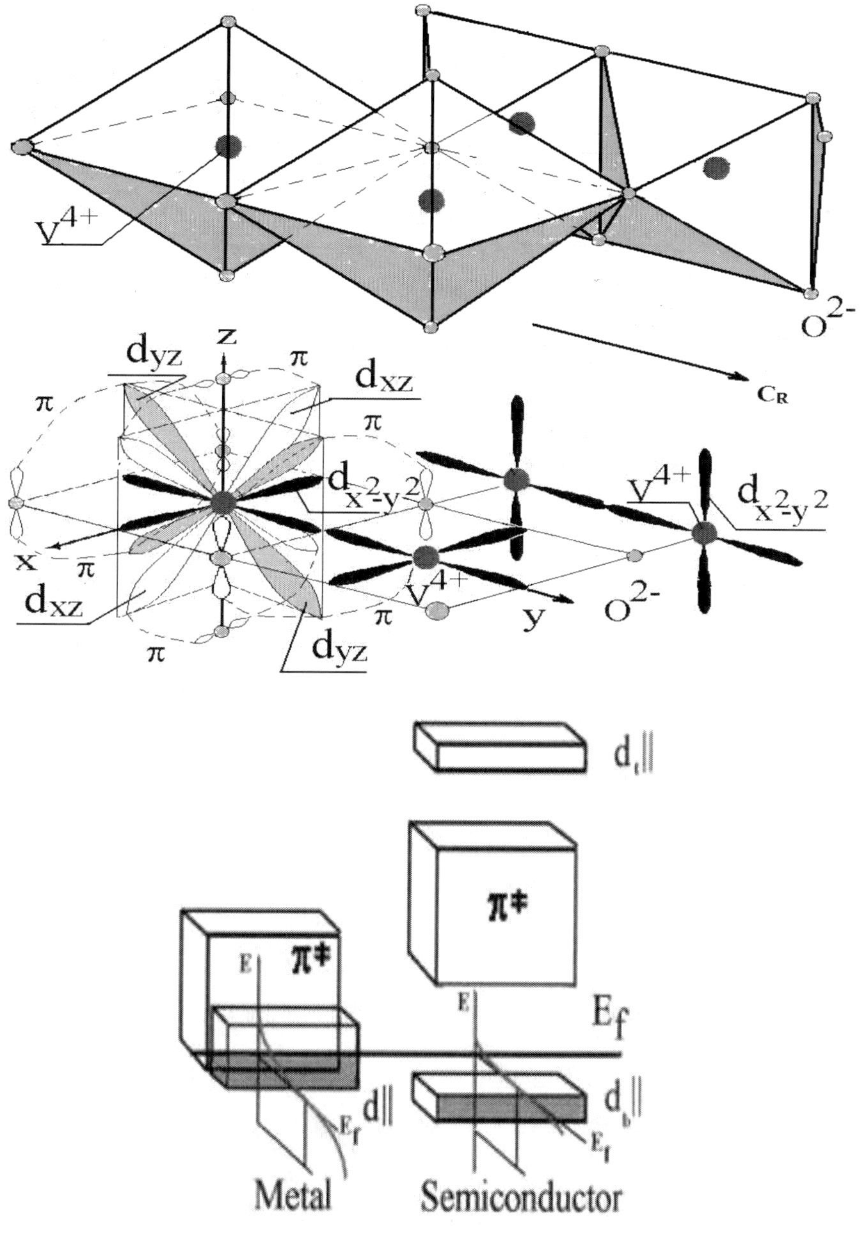

Figure 1. Schematic representation of the 3*d* orbitals of a V^{4+} ion in an oxygen octahedron ($T>T_c$). Energy band diagram of VO_2: the tetragonal metal phase at high temperature ($T>T_c$), the low temperature monoclinic semiconductor phase ($T<T_c$).

Thus, vanadium dioxide demonstrates a unique case, in which a 1D chain of hydrogen-like atoms, each with one electron at the *d*-orbital, is stabilized in space mainly by σ-bonds of oxygen octahedrons. However, in contrast to the electron in a hydrogen atom, this single electron is located, and on the *d*-orbital but not on the *s*-orbital. Unlike the spherically symmetric *s*-orbital, the *d*-orbital has cross-like branches, which can form 1D metal-type bonds with the orbitals of similar atoms of neighboring octahedrons arranged along the C_R axis.

For a chain of single electron atoms, the Schroedinger equation with the Hamiltonian that takes into account the interaction between only the nearest neighbors has an exact solution and can be written as [2]

$$[d^2\Psi(y)/dy^2] + (2m/\hbar^2)[E-V(y)]\Psi(y)=0.$$

Here, V(y) is the potential energy equal to the sum of the potential energies of an electron in the field of individual ions, without regard for a small addition, specifically, for the interaction energy of the electron of this ion with his neighboring ion (the direction of the y axis coincides with the direction of the rutile C_R axis (Figure 1). The solution of the Schroedinger equation predicts the existence of allowed and forbidden energy bands for free $3d^1$ electrons belonging to vanadium ions, which are members of the 1D chain. The electrons of $3d$ orbitals occupy the lowest states of the allowed $3d\|$ band, and according to the Pauli principle, this band is half filled with pairs of electrons with antiparallel spins. These electrons occupy the lower half of the band, since the number of levels in the band is equal to the number of members in the chain. The upper half of the allowed band remains unfilled, as happens in a typical metal; i.e., having one electron, the nonbonding orbitals of vanadium ions form a system of metal type bonds with neighboring ions of the chain. A specific feature of this system of bonds is that each atom supplies only one electron to the standard metal bond. In such a first approximation, metal-like conductivity along the 1D chain appears due to the half filling of the allowed electron band and is bound to persist at all temperatures, including low temperatures. Thus, such a system is prevented from undergoing any phase transformations.

Abandonment of the Rigid Chain Concept

In according to X-ray structural data [1], at $T < T_c$, the lattice symmetry lowers from the tetragonal to monoclinic type, since neighboring atoms in the chain approach in pairs to form stable ion pairs (dimers). Since there are two vanadium atoms in the VO_2 unit cell, the number of pairs is equal to the number of unit cells. The dimers appearing in chains distorts each unit cell and, as a result, distorts the entire crystal lattice, with the above-mentioned lowering of the lattice symmetry. Pairing of vanadium ions means that the spacing between vanadium ions within a pair (within a dimer) is smaller than the spacing between the pairs themselves; therefore, the lattice period along the chain doubles in comparison with the initial period. Because of such structural transformation, i.e., the Peierls transition, a band gap is formed in the electron spectrum [2], and the vanadium dioxide transforms into a semiconductor. The calculated band gap of such a semiconductor lies in the range 0.2–0.5 eV. In this case, the $3d\|$ band breaks into two subbands, each containing half of the total number of levels in the initial $3d\|$ band. According to the experiment [3], the splitting between the subbands is 2.5 eV, i.e., so large that we have to suppose that the π^* band is located between d_{x2-y2} – subbands, at an energy of 0.7 eV above the lower $3d\|$ subband. With the band gap of 2.5 eV, the lower $3d\|$ subband is almost completely filled with electrons, whereas the upper subband is practically empty. Therefore, in the model of a not rigid chain, the VO_2 crystal is a semiconductor with a 0.7 eV band gap between the lower 3*d*|| subband and the π* band (Figure 1), as reliably established experimentally.

The scheme modified in this direction can be improved further by taking into consideration the possibility of the thermal ionization of dimers, which allows vanadium dioxide to not remain a semiconductor at temperatures close to the PT temperature. In fact, at $T > 67°C$, vanadium dioxide is a metal, as follows from a large body of experimental data [1]. All of these data suggest that the lattice increases its symmetry from the monoclinic to the tetragonal type at $T > T_c$. However, simple estimates show that, with the 0. 7 eV band gap between the almost completely filled 3*d*|| subband and the practically empty π* band, increasing temperature to $T_c = 67°C$ cannot induce the complete destruction of dimers via the thermally activated transfer of electrons into the π* band. In addition, although the model for the doubling of the period of a 1D chain along the C_R axis (the Peierls model) allows an exact solution for VO_2, the model gives a value for the splitting between the two 3*d*|| subbands of no larger than 0.5 eV, whereas the splitting observed experimentally is 2.5 eV [3].

Consideration of Correlation Effects

Further development of the model of PT processes is based on the well-known Mott idea [4], consisting of the following: to solve the problem concerning the behavior of an electron in the field of a 1D chain of vanadium ions, the one particle approximation is inadequate, and it is essential to take into account the rather strong correlation interaction between electrons themselves. This interaction leads to dependence of the energy position of the bands on their occupation by electrons [5] and creates a correlation addition of ~2.0 eV to gap between the 3*d*|| and 3*d*|| subbands. In fact, even in the context of a simple model, in which a V–V dimer is represented as a hydrogen molecule, consideration for correlation interaction between electrons in the molecule yields the theoretical 3*d*||–3*d*|| splitting at ~2.5 eV. Thus, we obtain an additional increase in the splitting of the 3*d*|| band to ~2.5 eV due to correlation interactions; i.e., we are dealing with the formation of a wide gap semiconductor, whose band gap consists of the Mott gap of 2.0 eV and the Peierls gap of 0.5 eV. With this result, we finish the construction of a qualitative, but experimentally justified model of the band structure of the material's ground state. Obtaining a rigorous theoretical solution to a multiparticle problem is a rather complex challenge in itself. However, recent computer calculations in the context of the G0W0 + COHSEX approximation [5] show that, in the vicinity of the Fermi level E_f in the energy spectrum of 3*d* electrons of VO2, a 0.7 eV band gap is formed and the splitting of the 3*d*|| band is 2.5 eV. The same calculations show that the energy position of the bands depends on the degree of band filling with electrons, so that, as electrons are transferred to the unfilled bands, the energy gaps become narrower.

Thus, according to the multistage scheme developed specifically for the PT in vanadium dioxide, upon heating of the samples, correlation effects induce a decrease in the energy gap between the lower 3*d*|| subband and the π* band from the initial band gap of 0.7 eV, since the bands move in correlation with the thermally activated transfer of electrons from the lower 3*d*|| subband to the π* band (Figure 1). In this case, positive feedback between the band gap and the band populations is produced: a decrease in the band gap promotes an increase in the rate of the thermal generation of electrons into the π* band, and in turn, this increases narrows the band gap. It should be noted that, for the same reason, the upper $3d\|$ subband falls, but more slightly.

But simple estimations show that the thermal "tails" of Fermi distribution of electrons between energy levels are short (~0.030 eV) and are unable to provoke thermal decomposition of the dimers that is requisite for the structural PT. With this circumstance taken into consideration, we think that the mechanism for the semiconductor–metal PT is such as described below. With increasing temperature, correlation interaction between electrons narrows the energy gap between the bottom of the π^* band and the top of the $3d\|$ subband and they touch the Fermi level (Figure 1). This process initially proceeds without any noticeable breakage of the dimers. As the sample is heated further, the bottom of the π^* band lowers because correlations and do below the Fermi level, and as result, the transitions of electrons from the lower $3d\|$ subband to the π^* band are intensified; thus, the destruction of dimmers becomes more efficient. However, the complete decomposition of dimers, which is necessary for the structural transition to occur, only takes place, if the vast majority of electrons from the $3d\|$ subband forming the dimers transit to the π^* band. In turn, this is only possible if the number of states in the π^* band is substantially larger than that in the lower $3d\|$ subband, which in practice is indeed in the case for VO_2, since the number of levels in the lower $3d\|$ subband and the π^* band are, correspondingly, $N/2$ and $2N$, where N is the number of vanadium atoms. After the above-mentioned interaction, the $3d\|–\pi^*$ transitions are intensified in a narrow temperature region below the structural PT to the metal phase. Correlation interaction not only lowers the energy position of bands during their filling with electrons, but also extends "the tails" for electron distribution among the energy levels, thus reducing the step in the distribution function at the Fermi level (this step is commonly referred to as the Migdal jump [2]). The combined effect of the intensification of electron processes and the correlation-induced extension of the tails greatly facilitates the disruption of dimers by efficiently pumping electrons out from their σ bonds, which belong to the lower $3d\|$ subband. As the critical concentration of broken dimers is attained, the structural PT to the tetragonal phase proceeds abruptly. Thus, in this complex model of the PT, the Mott electron transition, at which the conductivity of the material gradually increases as a result of the correlation induced narrowing of the band gap, is primary and, upon heating of the sample, not only initiates the gradual approach of the material to the region of stimulated disruption of σ bonds in dimers, but also ensures the process of this accelerated breakage as such. It should be noted that, in itself, the onset of the accelerated breakage of dimers does not correspond to the instant of structural PT occurrence. Only at the instant when the critical concentration of broken dimers at the temperature $T = T_c + \Delta T$ is attained, a nucleus of the new tetragonal phase located within the old phase extends over the whole VO_2 film nanocrystal at the speed of sound, that reflecting the martensitic character of the PT in VO_2. (Here, ΔT is the deviation of temperature from T_c; the deviation is necessary for a viable overcritical sized martensitic nucleus of the new phase to be formed [6].) Thus, only at the point $T = T_c + \Delta T$ on the temperature scale, the material exhibits a stepwise structural PT to the new tetragonal metal phase under the influence of forces that are produced by σ-bonds of the oxygen octahedrons and vanadium ions that are returning to the centers of these octahedrons (Figure 1). It should be noted that, due to the structural PT, the upper $3d\|$ subband unevenly shifts down in energy scale only by the Peierls energy 0.5 eV because of the disappearance of doubling of period of the 1D chain. This means that the material transforms into the metal phase, with conduction occurring only over the π^* band; at the same time, to relation of the $3d\|$ band, the material remains a Mott insulator with a band gap of 2.0 eV. The doublet structure of the $3d\|$ band at this temperature is suggested by experimental data on the photoemission spectra of VO_2 excited by synchrotron radiation with the photon energy 60 eV [3]. The above proposed concept makes the prediction that, due to the sharply increased rate of correlation induced narrowing of the $3d\|$ band gap, the Peierls stepwise lowering of the upper $3d\|$ subband in turn is capable of stimulating the Mott insulator–metal electron transition now within the $3d\|$ band. In other words, the Peierls stepwise lowering of the upper $3d\|$ subband is capable of stimulating the decrease in the energy gap between the upper and lower $3d\|$ subbands from 2.0 eV to zero, as the material is heated further to temperatures above T_c. This is bound to yield extra metallization of the sample (Figure 1).

Thus, according to the described concept, the phase transformation proceeds in three stages: two stages occurring in a system of strongly correlated electrons and one stage in lattice system which increases of the crystal lattice symmetry by phase transformation [7-9]. This concept generates a possibility of creation of a wide set of new applied devices: smart windows, gas sensors, strain sensors, thermo-sensors, vibro-sensors, IR-bolometers, superbroadband transistors and superfast optical limiters.

REFERENCES

[1] W. Bruckner, H. Opperman, W. Reichelt, G. Wolf, F.A. Chudnovsky, E.I. Terukov. *Vanadium dioxide* (Berlin, Academy-Verlag, 1983).

[2] S. V. Vonsovskii and M. I. Katsnelson, *Quantum Solid State Physics* (Nauka, Moscow, 1983) [in Russian].

[3] S. Shin, S. Suga, M. Taniguchi, M. Fujisawa, H. Kanzaki, A. Fujimori, H. Daimon, Y. Ueda, K. Kosuge, and S. Kachi, *Phys. Rev*. B 41, 4993 (1990).

[4] N. F. Mott, *Metal–Insulator Transitions* (Nauka, Moscow, 1979; Taylor Francis, London, 1974).

[5] H. Jiang, R. I. Gomez_Abal, P. Rinke, and M. Scheffler, *Phys. Rev*. B 82, 045108 (2010).

[6] 6 A. L. Roytburd. UFN, 113, 69 (1974).

[7] A.V. Ilinskiy, O.E. Kvashenkina, E.B. Shadrin, *Semiconductors,* Vol. 45, No. 9, pp. 1153–1157 (2011).

[8] A.V. Ilinskii, V.Yu. Davydov, R.A. Kastro, O.E. Kvashenkina, M.E. Pashkevich, E.B. Shadrin Technical Physics Letters, Vol. 39, No. 8, pp. 705–708 (2013).

[9] I.N. Goncharuk, A.V. Ilinskiy, O.E. Kvashenkina, E.B. Shadrin, *Physics of the Solid State,* Vol. 55, No. 1, pp. 164–174 (2013).

PREPARATION AND INVESTIGATION OF POROUS SILICON NANOPARTICLES FOR TARGETED DRUG DELIVERY

Yu. M. Spivak*, E. V. Maraeva, A. O. Belorus, A. V. Molchanova, and N. R. Nigmadzyanova
St. Petersburg Electrotechnical University "LETI", Russian Federation

ABSTRACT

Porous silicon layers and nanoparticles were prepared in n-Si and characterized using a variety of analytical methods (atomic force microscopy (AFM), thermal desorption method). Measurements of the specific surface area were carried out by using the BET method showed that specific surface area of the por-Si powders varying from 50 m^2/g to 200 m_2/g; value of specific surface area is higher for por-Si powders, obtained from silicon with a lower resistivity. The study of size and morphology of Ni-based precipitates electrochemically deposited from aqueous solution of $NiCl_2$ on the substrate of Si and por-Si were carried out.

Keywords: porous silicon nanoparticles, electrochemical deposition of nickel, pore size distribution, atomic force microscopy

INTRODUCTION

Porous materials, including porous silicon (por-Si), obtained by electrochemical methods, are relevant for a wide range of different applications [1-3]. Area of application is largely determined by the parameters of the porous structure (pore geometry, their size distribution, porosity, specific surface area, phase composition, etc.). Nanoparticles based on porous silicon are very promising for applications in biomedicine [4-11]. For targeted drug delivery the porous silicon matrix is interesting as a carrier material for the delivery of molecular or nano payload (single or multiple drugs simultaneously) [12-15]. In the free volume that reaches more than 80% can be secured proteins, enzymes, drugs and genes. Also por-Si nanoparticles can be equipped with guides ligands for homing.

Overall, advantages of porous silicon as a material for the targeted delivery of drugs as follows: biocompatibility, i.e., a) the ability of the material incorporated into the body , to cause adverse clinical manifestations; b) ability to induce a cellular or tissue response necessary to achieve optimum therapeutic effect; biodegradation (dissolved and absorbed); the relative simplicity of obtaining technology; handling properties (variation of properties (porosity, surface development, hydrophobicity/hydrophilicity, the local distribution of

*Corresponding author: PhD. Yulia M. Spivak. Docent, Department of micro- and nanoelectronics, Saint Petersburg Electrotechnical University "LETI", 197376, Russian Federation, Saint-Petersburg, Prof. Popova street. 5. Russian Federation. Tel.: + 7 812 234 3164. E-mail: ymkanageeva@yandex.ru.

surface charges , etc.) in a wide range); multifunctionality (not only the carrier matrix, but simultaneously the sensor mark , etc.).

Properties of por-Si (optical, electrical, mechanical, etc.), parameters of the porous structure (porosity, specific surface area, pore sizes, pore geometry and structure, etc.) will be determined by choice of the parameters raw material, typically monocrystalline silicon, as well as the process conditions of the electrochemical etching [1, 16-19]. Now quite an extensive knowledge about the influence of technological conditions upon the characteristics of porous silicon structure are accumulated. However, it is necessary to develop preparation techniques for nanoparticles and nanopowdes from por-Si, as well as ways of their separation by size, the choice of storage conditions and so on. Also important to study the influence of porous silicon nanoparticles (especially modified by other substances) from the viewpoint of safety and interaction with living organisms [20].

Moreover, the introduction of magnetic materials into the porous silicon matrix is promising in terms of extending the functionality of the nanoparticles. To achieve this, the method appears promising introduction of magnetic material by electrochemical deposition into the por-Si layer, followed by grinding to obtain the nanoparticles.

The purpose of this paper was to study the influence of the silicon substrate parameters upon the specific surface area and morphology of porous silicon nanoparticles and Ni-modified por-Si layers.

MATERIALS AND METHODS

Layers of porous silicon were formed in n-Si (111) (0.3 and 1.0 Ω·cm resistivity) and p-Si (100) (10 Ω·cm resistivity). Electrochemical etching is carried out for 25 minutes at a current density of 70 mA/cm2 anodizing in galvanostatic mode. The electrolyte based on an aqueous solution of hydrofluoric acid with the addition of isopropyl alcohol and hydrogen peroxide was used. Next, using ultrasonic treatment porous layer separated and milled. The resulting dispersions were dried, thereby to obtain powders.

Electrochemical (cathodic) deposition of nickel was performed from aqueous solution of $NiCl_2$ on the substrate of Si and por-Si.

The investigations of porous silicon morphology were carried out by atomic force microscopy (AFM, Ntegra Terrna, NT-MDT). For AFM studies the por-Si nanoparticles were deposited on glass substrates and dried. Measurements of the specific surface area were carried out by using the BET method on the equipment Sorbi® N.4.1. [21]. Experimentally selected optimal conditions for preliminary heat treatment of the samples were found (temperature, time).

RESULTS AND DISCUSSION

It is shown that por-Si powders of different types of these conditions are different. It is shown that specific surface area of the por-Si powders varying from 50 m2/g to 200 m2/g; value of specific surface area is higher for por-Si powders, obtained from silicon with a lower resistivity.

The pore size distribution was obtained by the full adsorption/desorption isotherms of capillary condensation method for the por-Si powder, obtained from n-Si (resistivity of 1 Ω·cm). It was found the presence of 34.5% of pores with a size of 3 to 5 nm, 7.2 % with pore sizes ranging from 5 to 6 nm, and 14% pores with a size of 6 to 8 nm, 19.8% with pore sizes ranging from 8 to 11 nm, 5.4% with pore sizes from 15 to 25 nm 19.1 % with pore sizes from 40 to 70 nm.

It was found that increasing the voltage increases the deposition of size and the density of nucleation. Increasing the deposition time, and increasing concentration of the solution leads to an increase in the size of nuclei and their concentration.

REFERENCES

[1] A.S. Lenshin, V.M. Kashkarov, Yu. M. Spivak, V.A. Moshnikov. Investigations of nanoreactors on the basis of p-type porous silicon: Electron structure and phase composition// *Materials Chemistry and Physics*, Volume 135, Issues 2–3, 15 August 2012, Pages 293-297.

[2] V. A. Moshnikov, I. E. Gracheva, A. S. Lenshin, Yu. M. Spivak, M. G. Anchkov, V. V. Kuznetsov, J. M. Olchowik. Porous silicon with embedded metal oxides for gas sensing applications // *Journal of Non-Crystalline Solids*, vol. 358, is. 3, 1 February 2012, Pages 590–595.

[3] T. M. Zimina, E. N. Muratova, Yu. M. Spivak, V. E. Drozd, A.A. Romanov. Tekhnologii formirovaniya i primenenie nanosloev i nanoporistykh kompozitsii Al_2O_3 dlya micro- nanotekhniki // *Nano- i microsystemnaya tekhnika*, 2012. № 12. PP. 15-24. (in russian).

[4] Raul J. Martin-Palma, Miguel Manso-Silvan; Vicente Torres-Costa. Biomedical applications of nanostructured porous silicon: a review // *J.Nanophoton.*2010. 4(1), 042502.

[5] B. Gupta,Y. Zhu,B. Guan, etc. Functionalised porous silicon as a biosensor: emphasis on monitoring cells in vivo and in vitro // *Analyst,* 2013,138, 3593-3615.

[6] T. J. Barnes, Karyn L J., C. A. Prestidge.Recent advances in porous silicon technology for drug delivery // *Therapeutic Delivery*, 2013. Vol. 4, No. 7. PP. 811-823.

[7] Ksenofontova O. I., Vasin A. V., Egorov V. V., etc. Poristyi kremnii i ego primenenie v biologii i meditsine // *JTF*, 2014. T. 84, is. 1. PP. 67-77. (in russian)

[8] Porous silicon in drug delivery devices and materials / E. J. Anglin, L.C. Cheng, W. R. Freeman, and all// *Adv Drug Deliv Rev*. – 2008. – Vol. 60. – P.1-32.

[9] Iskusnykh I. Yu., Popov A. L., Popova T. N., etc. Vliyanie nanokrystallicheskogo kremniya na metabolicheskuyu aktivnost i proliferatsiyu fibroplastov i kletok kartsinomy gortani // Vestnik VGU, ser. Khimiya, Biologia, Farmatsiya, 2012. № 1. PP. 96-102. (in russian).

[10] H. A. Santos, L. M. Bimbo, Barbara Herranz, etc. Nanostructured porous silicon in preclinical imaging: Moving from bench to bedside // *Journal of Materials Research,* 2013. Vol. 28.Is. 02. PP. 152-164.

[11] S. Dhanekar, S. Jain, J. M. Islamia, etc. Porous silicon biosensor: Current status // *Biosensors and Bioelectronics*, 2013. Vol 41.Pages 54–64.

[12] D.Liu, L. M. Bimbo, E. Mäkilä, etc. Co-delivery of a hydrophobic small molecule and a hydrophilic peptide by porous silicon nanoparticles // *Journal of Controlled Release,* 2013. Vol. 170,is. 2. Pages 268–278.

[13] E. Secret, K. Smith, V. Dubljevic, etc. Antibody-Functionalized Porous Silicon Nanoparticles for Vectorization of Hydrophobic Drugs // *Advanced Healthcare Materials,* 2013. Volume 2, Issue 5, pages 718–727.

[14] W. Xu, J. Riikonen,V.-P. Lehto.Mesoporous systems for poorly soluble drugs // *International Journal of Pharmaceutics,* 2013. Vol.453,is. 1. PP. 181–197.

[15] W. Xu, J. Riikonen,V.-P. Lehto. Mesoporous systems for poorly soluble drugs // *International Journal of Pharmaceutics*, 2013. Vol.453,is. 1. PP. 181–197.

[16] A. S. Lenshin, V. M. Kashkarov, P. V. Seredin, Yu. M. Spivak and V. A. Moshnikov. XANES and IR Spectroscopy Study of the Electronic Structure and Chemical Composition of Porous Silicon on n- and p-Type Substrates // *Semiconductors*, 2011. Vol. 45, No. 9. PP. 1183-1188.

[17] A. S. Lenshin, V. M. Kashkarov, Yu. M. Spivak, V.A. Moshnikov. Study of Electron Structure and Phase Composition of Porous Silicon // *Glass physics and chemistry,* 2012. Vol. 38, No. 3. PP. 315-321.

[18] Spivak Y. M., Moshnikov V. A., Kuznetsov, etc. Self-Organized Porous Structure with Several Levels of Pores in Electrochemical Anodized Silicon // 6th Workshop on Functional and Nanostructured Materials, *10th Conference on Intermolecular and Magnetic Interactions in Matter*, 27–30 September 2009, Sulmona–L'Aquila, Italy, 2009, P067.

[19] A. S. Lenshin, V. M. Kashkarov, D. L. Goloshchapov, etc. Sostav i reactsionnaya sposobnost nanoporoshkov poristogo kremniya // *Neorganicheskie Materialy*, 2012. T. 48, is. 10. PP. 1–6. (in russian).

[20] Mohammad-Ali Shahbazi, Mehrdad Hamidi, Ermei M. Mäkilä, etc. The mechanisms of surface chemistry effects of mesoporous silicon nanoparticles on immunotoxicity and biocompatibility // *Biomaterials*, 2013. Vol. 34, is. 31. PP. 7776–7789.

[21] A. S. Lenshin, E. V. Maraeva. Issledovanie udelnoy poverkhnosti perspektivnykh poristykh materialov i nanostruktur metodom teplovoi desorbtsii // *Izvestiya SPbGETU* "LETI". 2011. № 6. PP. 9-16. (in russian).

ACID-BASIC ASPECT OF CONTROL OF NANOCOMPOSITE ELECTRIC PROPERTIES

*Maxim Sychov**

St. Petersburg State Institute of Technology (Technical University)
St. Petersburg, Russia

Polymer-inorganic composites are used in electronic devices, printed electronics, embedded electronics applications etc. Properties of composites, including electric ones, depend on composite structure, Figure 1.

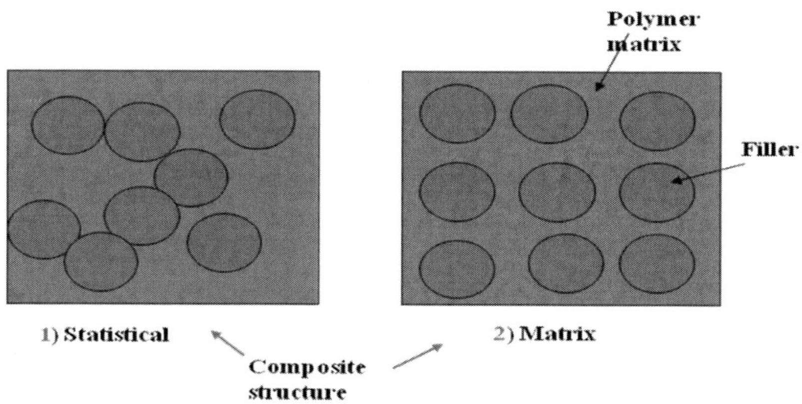

Figure 1. Composite structures.

Composite structure vary depending on interfacial interactions, namely depending on intensity of particle-particle, particle-polymer and polymer-polymer interactions. Such interactions may have acid-basic nature. Below examples of such considerations are given for several composites. In a first example dielectric composite is described comprising poly(vinyl alcohol) cyanethyl ester (CEPVA) as a polymer binder with the barium titanate ($BaTiO_3$) as a filler. For dielectric composites important characteristic is dielectric constant. Concentration dependences of dielectric constant of composites ε were fitted with Lichtenecker equation:

$$\varepsilon^k = \varphi_1 \varepsilon_1{}^k + \varphi_2 \varepsilon_2{}^k \tag{1}$$

where ε_1, ε_2 – dielectric constants of polymer and filer, φ_1, φ_2 – their volume fractions; k – constant.

Surface acid-base properties of filler were changed by various methods: heat-treatment, hydrolysis, utilization of surfactant etc. [1]. After that acid-base properties of filler surface was studied by adsorption of indicator molecules [2]. In a number of experiments it is shown

*msychov@yahoo.com.

that dielectric constant of composite depends on amount of Broensted basic centers on surface of filler due to their interaction with acidic hydroxyls of polymer CEPVA [3,4]. For approximation of obtained results a modified Lichtenecker equation is introduced which gives good agreement between the experimental and calculated $\boldsymbol{\varepsilon}$ values:

$$\varepsilon^{m1+m2*BB} = \varphi_1 \varepsilon_1^{m1+m2*BB} + \varphi_2 \varepsilon_2^{m1+m2*BB} \quad (2)$$

where ***BB*** – amount of Broensted basic centers on surface of filler capable of interaction with acidic hydroxyls of polymer CEPVA, ***m1*** and ***m2*** are fit coefficients. Example of fit of experimental data with equation (2) is given in Figure 2.

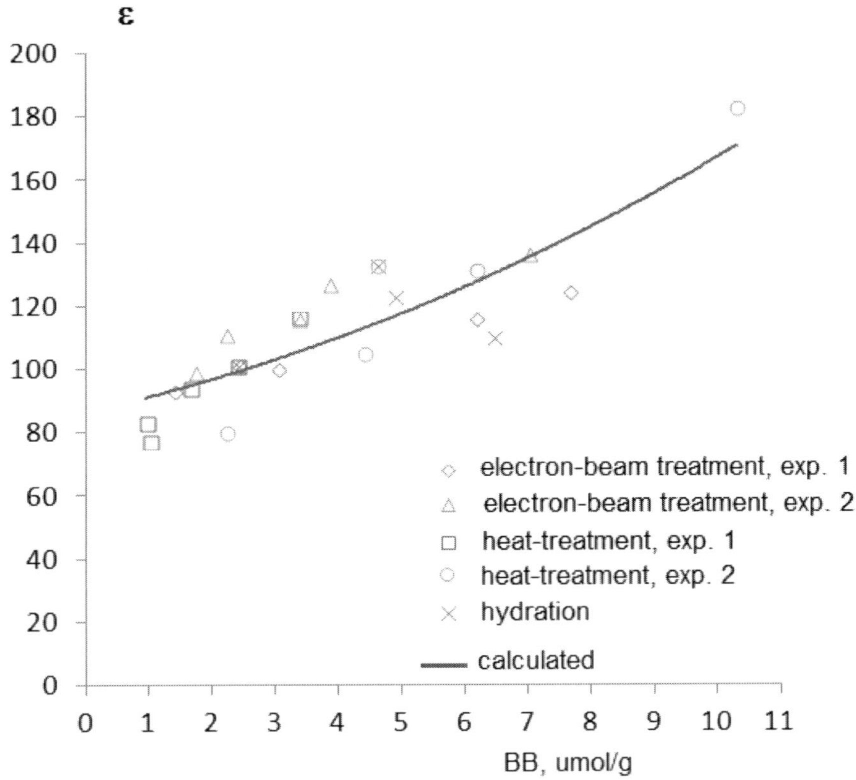

Figure 2. Example of fit of experimental data with equation (2).

Further experiments were continued using other methods of modification of filler surface, such as core-shell (treatment made by Prof. O.A. Shilova), plasma (treatment made by Dr. A.A. Eruzin). Experiments confirmed validity of approach and equation (2). This equation is unique in having prognostic power to predict electric properties of composites depending on interfacial interactions. It should be noted that depending on polymer matrix, content of other active centers may be utilized in equation (2) depending on peculiarities of polymer – solid interactions. Validity of approach was proved using several fillers and polymers such as butadiene nitrile rubbers, acrylic resin.

In case of latex (water dispersion of polymer) utilization of content of specific surface centers is not suitable due to complexity of the system – latex contains surfactants etc. In this case good results are obtained by introduction of new parameter

$$\Delta = pH_{latex} - H_o, \quad (3)$$

where pH_{latex} is pH of latex, H_o – acidity of filler's surface. In this case parameter Δ is used instead of parameter BB in the equation (2). Validity of approach was shown using several fillers (BaTiO3, ceramics VS-1) and latexes (butadiene nitrile BN-30K-2, acrylate BMNK 60/1 and DMMA-65GP), Figure 3.

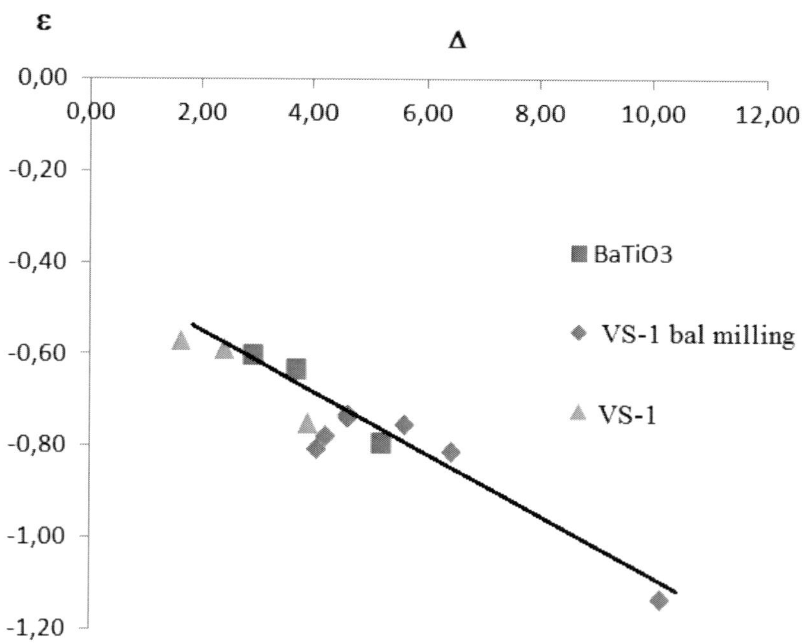

Figure 3. Interrelation between Δ and ε for latex composites.

Further using butadiene nitrile latex BN-30K-2 and conductive fillers (carbon blacks K354 and PM-80E and graphite S-1) it was shown that parameter Δ also is suitable in prognosis of electric conductivity of composites. Concentration dependencies of conductivity were approximated by equation of percolation theory [5]:

$$\sigma = \sigma_0 (c - c_n)^t , \quad (4)$$

where c – content of conductive filler, % vol.; c_n – percolation threshold, %vol.; t — index; σ_0 — constant. As it may be seen from Table 1, value of percolation threshold is in reverse correlation with value of parameter Δ. Consequently value of conductivity is in direct correlation with value of parameterΔ.

Table 1. Properties of conductive composites

Filler	Δ	$c_{n,}$ % vol.	σ, Sm/m
S-1	7,9	15,6	50
K354	4,2	23	0,22
PM-80E	8,3	7,7	250

Thus, in a number of experiments it is shown that consideration of acid-basic properties of surface of filler and acid-basic interface interactions is a productive approach in explanation and prediction of composite electric properties. It may be expected that other properties sensible to such interactions in composite may be also predicted and promoted by consideration of acid-basic aspect control of acid-basic properties of surface of filler, for example composite mechanical properties.

REFERENCES

[1] S. A. Alekseev, V.G. Korsakov, M.M. Sychov, O.V. Likhacheva, A.G. Rodionov, L.L. Ezhenkova, Effect of Donor–Acceptor Sites at a Barium Titanate Surface on the Properties of Composites Based on Cyanoethyl Polyvinyl Alcohol. *Rus. J. Appl. Chem.*, 2006, Vol. 80, No. 4, pp. 700–703.

[2] M. M. Sychov, V.V. Bakhmet'ev, Y. Nakanishi, S.V. Mjakin, L.V. Havanova, O.A. Cheremisina, V.G. Korsakov. Surface properties of ZnS and AC powder electroluminescent phosphors. *Journal of the SID* 11 (1), p.33-38 (2003).

[3] Sychov M. et al. // *Glass Physics and Chemistry*. 2011. V.37. No.5. P.734.

[4] Sychov M. et al. // *Glass Physics and Chemistry*. 2011. V.37. No.6. P.624.

[5] Aneli, J.N. *Structuring and conductivity of polymer composites* / J.N. Aneli, L.M. Khananasvili, G.E. Zaikov. – NY : Nova science publishers Inc., 2007. – P.37.

CONTROL OF DIELECTRIC PROPERTIES FOR POLYMER-INORGANIC NANOCOMPOSITES VIA THE MODIFICATION AND MODELLING OF INTERFACIAL INTERACTIONS INVOLVING SURFACE FUNCTIONAL GROUPS

Maxim M. Sychov*[*,1], Ekaterina S. Vasina[1], Sergey V. Mjakin[1], Alexander A. Eruzin[1], Tamara V. Khamova[2], and Olga A. Shilova[2]

[1]St. Petersburg State Institute of Technology (Technical University)
St. Petersburg, Russia
[2]Institute of Silicate Chemistry of Russian Academy of Sciences
St. Petersburg, Russia

INTRODUCTION

Hybrid polymer-inorganic composites are unique compounds featuring with a unique combination of elastic and adhesive properties of plastics with hardness and other specific characteristics of solid fillers. Particularly, materials of this kind are useful as dielectric barrier layers in capacitors, displays, electroluminescent panels and other electronic devices. The main requirements to such composites include a high permittivity (k) and low dielectric losses ($tg\delta$).

An essential factor responsible for the properties of polymer-inorganic composites is the interaction between their components that is in turn determined by the filler surface properties. A promising approach to the improvement of hybrid composite performances is based on the control over the functional composition of the filler surface. Earlier [1,2] we synthesized and studied composites comprising barium titanate ($BaTiO_3$) as a filler with extremely high k value and cyano-ethyl ester of poly(vinyl alcohol) (CEPVA) as a polymer binder with one of the highest k values among the polymers due to the presence of highly polar C≡N, C=O and OH groups capable of orientation in electric fields. The dielectric properties of these materials were found to effectively controllable via $BaTiO_3$ surface modification. In the continuation of these studies, the control over the dielectric properties of the considered composites by the filler surface modification, possible mechanism of surface functional transformations and simulation of their effect upon the dielectric properties of the composites are discussed in this paper.

EXPERIMENTAL

Barium titanate HPBT-1B with particle size 300-400 nm and k ~ 4400 produced by FujiTitanium (Japan) was modified using the following methods:

*msyhov@yahoo.com.

1. Hydration in distilled water.
2. Sol-gel deposition of SiO_2 (by tetraethoxy-silane (TEOS) hydrolysis) as well as SiO_2:Al_2O_3 (70:30 %wt.), SiO_2:polyionene PI (oligomeric salt of quarternary ammonia), SiO_2:hyperbranched polymer HBP (four-branched polyol with 64 end OH-groups) mixtures from TEOS with the corresponding additives and Ta_2O_5 (by $TaCl_5$ hydrolysis) according to the process generally described in [2] followed by annealing at 700°C within 1 hour to remove the residual non-reacted components.
3. Plasma deposition of MgO_x and ZrO_2 layers using a Bulat-NNV-6.6I-1 installation.

The amount of all the additives deposited onto $BaTiO_3$ surface was about 1% wt. The modified $BaTiO_3$ samples were characterized by measuring their specific surface according to BET method (nitrogen desorption using a SORBI®N.4.1 device) and study of surface functional groups using the adsorption of acid-base indicators with intrinsic pK_a values in the range from –5 to 15 described in detail in [2]. The obtained materials were dispersed in 30%wt. CEPVA (Plastpolymer Co. St-Petersburg, $M_w \approx 55000$, ratio [CN]:[OH]≈57:43, $k \approx 19$) solution in DMF so that the filler content in the target dry materials be 40%vol. The composites were deposited onto aluminum supports and dried followed by measuring their dielectric properties using the device E7-20 as described in [2].

RESULTS AND DISCUSSION

The data summarized in Table 1 show that SiO_2 deposition onto the filer surface provided an almost double k increase compared with the composite based on non-modified $BaTiO_3$in combination with reduced $tg\delta$. On the contrary, for all other materials k value markedly dropped.

Table 1. Properties of modified $BaTiO_3$ samples and CEPVA-$BaTiO_3$ composites

Parameter			S,m^2/g	k	$tg\delta$
Non-modified $BaTiO_3$			2.1	125.8	0.36
$BaTiO_3$ modified using the following methods		Hydration	1.7	68.4	1.30
	Sol-gel deposition	SiO_2	188.0	242.3	0.22
		SiO_2+Al_2O_3	8.1	81.3	0.80
		SiO_2+PI	31.2	86.8	0.05
		SiO_2+HBP	38.5	38.9	0.11
		Ta_2O_5	10.7	81.0	0.12
	Plasma deposition	MgO	-	55.6	0.18
		ZrO_2	-	44.0	0.29

The observed effect can be partially attributed to a drastic increase of the specific surface for SiO_2–modified filler resulting in the enhanced interaction with the polymer binder. However, no general correlation between the filler specific surface area and permittivity was found that suggests that a specific role of certain active centers rather that the effect of the whole surface. The comparison of the filler surface functional composition and dielectric properties of the composites allowed us to establish the following correlations:

1. Permittivity of the composites grows with the content of Brensted neutral (pKa ~7.3) and basic (pKa ~12.8) centers (hydroxyls) on the filler surface (Figure 1) due to their ability to react with acidic hydroxyls in CEPVA thus improving the compatibility in the system.

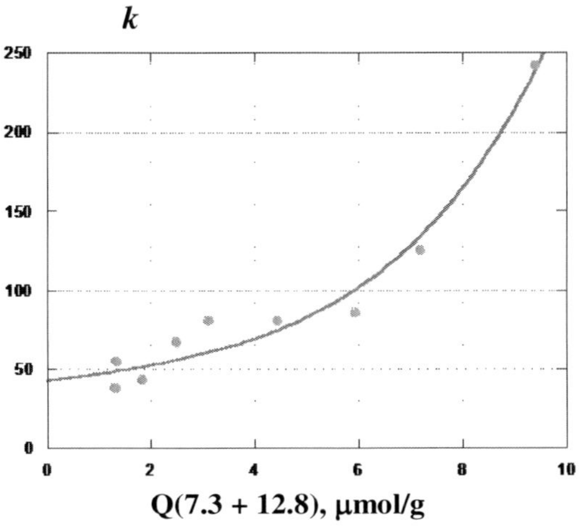

Figure 1. Permittivity of CEPVA-$BaTiO_3$ composites vs the content of surface centers with pKa 7.3 and 12.8 approximated by the Likhtenekker eqn.

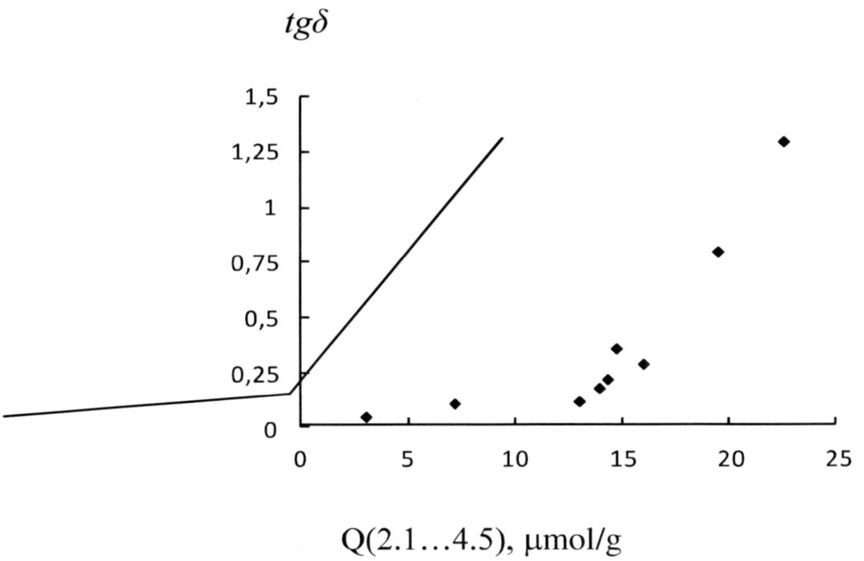

Figure 2. *tgδ* of CEPVA-$BaTiO_3$ composites vs the content of surface centers with pKa 2.1…4.5.

This trend is well approximated by a Lichtenecker equation [3]:

$$k^{\alpha} = \varphi_1 k_1^{\alpha} + \varphi_2 k_2^{\alpha}$$

where k_1 and k_2 are permittivity values (4400 and 19 correspondingly), φ_1 and φ_2 are their volume parts (0.4 and 0.6), $\alpha = m_1 + m_{2*} Q_{(7,3+12,8)} = -1...1$ - coefficient taking into account the concentration of the considered functional groups; m_1 and m_2 – coefficients.

The value tgδ of the composites inversely correlates with the content of Brensted acidic centers with pK_a 2.5...4.1 (Figure 2). Two almost linear regions in this plot probably determined by water adsorption forming the first and subsequent layers correspondingly are modeled in a good agreement with experimental data.

Conclusion

The considered approach based on the filler surface functionalization with specific functional groups and modeling of their interaction with polymer binder is promising for the development of hybrid polymer-inorganic composites with enhanced and adjustable dielectric performances and other useful properties.

References

[1] S.A.Alexeev, M.M.Sychov, O.V.Likhacheva, V.G.Korsakov, A.G.Rodionov, L.L Ejenkova. High dielectric constant composites for the electroluminescent displays. *Abs. book of the Display Optics'04 International Seminar.* St.Petersburg, October 18-20, 2004, p.35.

[2] Mjakin, S.V., Korsakov, V.G., Panova, T.I., Sosnov, E.A., Fomchenkova, Yu.S., Sychov, M.M., and Shilova, O.A., 2011, Effect of the modification of barium titanate on the permittivity of its composites with cyanoethyl ester of polyvinyl Alcohol, *Glass Physics and Chemistry*, 37, 624-628.

[3] New materials, Edited by Ju. S. Karabasov, 2002, MISIS Publishers, Moscow.

SYNTHESIS AND CHARACTERIZATION OF Y2O3:EU3+ NANOSIZED PHOSPHORS FOR PHOTODYNAMIC THERAPY OF CANCER

Anna B. Vlasenko, Lev A. Lebedev, Vadim V. Bakhmetyev, Maxim M. Cychov and Sergry P. Bogdanov

Department of Material Science,

St. Petersburg State Institute of Technology (Technical University), Russia

Objectives: development of technology for the synthesis of nanosized phosphors for improvement techniques photodynamic therapy of cancer.

Object of research: $Y_2O_3:Eu^{3+}$ nanosized phosphors.

Photodynamic therapy (PDT) is a promising modern approach to the treatment of cancer. PDT method is based on the introduction of a photosensitizer (compound capable of selective accumulated in a cancer locus and photo-stimulated yield of oxygen to destroy the tumor) into the organism [1]. However, PDT application is restricted by the complications related to light transmission to inner tumors. A possible approach to this problem is based on the development of special medications involving a photosensitizer in combination with a nanosized phosphor converting X-ray radiation (easily penetrating inside the organism) into the visible light with the wavelengths required for the photosenitizer activation.

The main requirements to the target nanosized phosphors include non-toxicity, biocompatibility, hydrolytic stability in blood, particle size less than 100 nm affording the formation of stable colloids and efficient generation of visible light with the required wavelength upon stimulation with “hard” X-rays commonly used in medicine. A possible compound useful as such a nanosized phosphor is $Y_2O_3:Eu^{3+}$. This study is aimed at the development of a process for obtaining finely dispersed $Y_2O_3:Eu^{3+}$ phosphors and characterization of their physicochemical and luminescent properties.

The dispersed $Y_2O_3:Eu^{3+}$ phosphors were prepared using the following two approaches:

- Pechini method based on a thermal decomposition of gels obtained by etherification involving citric acid complexes, metal ions and ethylene glycol to yield a polymer containing uniformly distributed yttrium and europium ions [2];
- self-propagating high temperature synthesis (SPHTS) based on thermally initiated reactions in a mixture containing yttrium nitrate and europium nitrate (oxidizer) in combination with glycine (“fuel”) flaming to form a solid foam comprising nanosized particles [3].

Thus synthesized phosphors were annealed at 700 or 800 °C. Europium content if the prepared samples varied from 2.5 to 12 % wt.

Annealing at 700°C resulted in obtaining phosphors with a very poor photo- and X-ray stimulated luminescence. On the contrary, the samples annealed at 800 °C are featured with an efficient luminescnence upon excitation with UV (λ_{max} = 365 nm) and “hard” X-ray (λ = 0.124…0.310 Å) irradiation.

The samples prepared using SPHTS procedure are featured with smaller particles compared with the phosphors obtained by Pechini process. However, the phosphors synthesized according to Pechini method provide almost twice higher brightness. Photo- and X-ray luminescence spectra of these two series of samples are almost the same with the most intensive emission band at 612 nm. The highest UV- and “hard” X-ray-stimulated luminescence intensities are observed for the samples containing 12 and 10 % wt. Eu correspondingly.

XRD data for $Y_2O_3:Eu^{3+}$ samples prepared by Pechini method are shown in table (1).

Table 1. XRD data for $Y_2O_3:Eu^{3+}$ samples prepared by Pechini method

Annealing temperature, °C	Eu content, % wt.	Lattice constant ***a***, nm	Crystallite size, nm	Lattice micro-strains $\Delta a/a$
700	7	1.0607	6.00	-0.0186
	10	1.0603	6.60	-0.0151
	12	1.0604	7.25	-0.0129
800	7	1.0613	10.9	-0.0077
	10	1.0600	13.1	-0.0060
	12	1.0611	15.2	-0.0050

The data in table (1) indicate that the increase of annealing temperature leads to the growth of crystallite size. The increase of Eu content in the phosphor also results in the increase of crystallite size in combination with the decrease of micro-strains $\Delta a/a$ in the crystal lattice. These micro-strains evidently correlate with the crystallite size since large crystallites are featured with lower surface responsible for numerous defects. Interestingly, the highest X-ray induced luminescence intensity is observed for $Y_2O_3:Eu^{3+}$ sample containing 10% wt. Eu with the lowest lattice constant ***a***. The lowest FWHM of XRD peaks for this sample suggests that the incorporation of 10% wt. Eu provides phosphors with the most perfect and least distorted crystal lattice with the lowest content of defects quenching the luminescence.

CONCLUSION

We synthesized phosphors $Y_2O_3:Eu^{3+}$ with nanosized crystallites having an intense “hard” X-ray-stimulated luminescence that suitable to create a pharmacological preparation for photodynamic therapy of cancer. The influence of the synthesis method, the annealing temperature and the Eu concentration on the crystal structure of obtained phosphors was investigated.

REFERENCES

[1] Dougherty T.J., Gomer C.J., Henderson B.W., Jori G., Kessel D., Korbelik M., Moan J., Peng Q. *Photodynamic Therapy. J. National Cancer Inst.,* 90(12), (1998) 889-905.

[2] Kodaira C.A., Stefani R., Maia A.S., Felinto M.C.F.C., Brito H.F. Optical investigation of $Y_2O_3:Sm^{3+}$ nanophosphor prepared by combustion and Pechini methods. *J. of Luminescence,* 127, (2007) 616-622.

[3] Balabanov S.S., Bykov Yu.V., Egorov S.V., Eremeev A.G., Gavrishchuk E.M., Khazanov E.A., Mukhin I.B., Palashov O.V., Permin D.A., Zelenogorsky V.V. Transparent Yb:$(YLa)_2O_3$ ceramics produced by self-propagating high-temperature synthesis and microwave sintering. *Optical Materials*, 35, (2013) 727-730.

INDEX

A

B

C

D

E

O

P

Q

R

U

V

W

X

Y

Z